Heterojunction Solar Cells (a-Si/c-Si)

Investigations on PECV Deposited Hydrogenated Silicon Alloys for use as High-Quality Surface Passivation and Emitter/BSF

A

DISSERTATION

submitted in partial satisfaction
of the requirements for the degree of

DOKTOR-INGENIEUR (DR.-ING.)

at the
Department of Mathematics and Information Technology
of the
University of Hagen, Germany

THOMAS MUELLER (DIPL.-ING.)

Hagen 2009

Bibliografische Information der Deutschen Nationalbibliothek

Die Deutsche Nationalbibliothek verzeichnet diese Publikation in der Deutschen Nationalbibliografie; detaillierte bibliografische Daten sind im Internet über http://dnb.d-nb.de abrufbar.

ISBN 978-3-8325-2291-9

Logos Verlag Berlin GmbH
Comeniushof, Gubener Str. 47,
10243 Berlin
Tel.: +49 030 42 85 10 90
Fax: +49 030 42 85 10 92
INTERNET: http://www.logos-verlag.de

Cover Image Credit: *NASA*

Submission date: January 27, 2009
Disputation date: April 3, 2009

Committee in charge:

Prof. Dr. Wolfgang R. Fahrner, University of Hagen, Germany
Prof. Dr. Heinrich C. Neitzert, University of Salerno, Italy

A new scientific truth does
not triumph by convincing
its opponents and making
them see the light, but
rather because its
opponents eventually die,
and a new generation grows
up that is familiar with it.

(Max Planck)

ABSTRACT

The main focus of the present work is related to the optimization of hetero-junction solar cells. The key role in obtaining high efficient heterojunction solar cells are mainly the PECV deposition of very low defect layers, and the sufficient surface passivation of all interfaces the heterostructure cell device consists of. In heterojunction solar cells, the a-Si:H/c-Si hetero-interface is of significant importance, since the hetero-interface characteristics directly affect the junction properties and thus solar cell efficiency. In this work, the deposition and film properties of various hydrogenated amorphous silicon alloys, such as a-SiC:H, a-SiO$_x$:H, and μc-Si:H (standard a-Si:H(i) is used as reference), are employed. Special attention is paid to (i) the front and back surface passivation of the bulk material by high-quality wide-gap amorphous silicon suboxides (a-SiO$_x$:H), and (ii) the influence of wide-gap high-quality a-Si:H-based alloys at the front side for use as emitter and passivation to suppress absorption losses.

(i) Surface Passivation by plasma deposited high-quality a-SiO$_x$:H. In heterojunction solar cells, the a-Si:H/c-Si hetero-interface is of significant importance, since the hetero-interface characteristics directly affect the junction properties and thus solar cell efficiency. A novel passivation scheme, featuring PECV deposited hydrogenated amorphous silicon suboxide films (a-SiO$_x$:H), has been investigated extensively. The a-SiO$_x$:H films are deposited by decomposition of silane (SiH$_4$), carbon dioxide (CO$_2$) and hydrogen (H$_2$) as source gases. Those films exhibit a significantly lower blue light absorption than standard a-Si:H(i), enabling a thicker passivation layer than standard a-Si:H(i) and therefore an increasing passivation quality. The impact of post-annealing at low temperatures of those films shows a beneficial effect in form of a drastic increase of the effective lifetime. This improvement of the passivation quality by low temperature annealing for the a-SiO$_x$:H likely originates from defect reduction of the film close to the interface. Raman spectra reveal the existence of Si-(OH)$_x$ and Si-O-Si bonds after thermal annealing of the layers, leading to a higher effective lifetime, as it reduces the defect absorption of the sub-oxides. Record high effective lifetime values of *4.7 ms on 1 Ωcm n-type FZ wafers* and 14.2 ms on 130 Ωcm p-type FZ wafers prove the applicability for a surface passivation of silicon wafers applicable to any kind of silicon based solar cells. These values appear to be the highest ever reported on a high doped crystalline wafer of 1 Ωcm resistivity. Samples prepared in this way feature a high quality passivation yielding effective lifetime values exceeding those of record SiO$_2$ and SiN$_x$ passivation schemes. The use of a-SiO$_x$:H may be a promising alternative for any passivation scheme existing.

(ii) Investigations on wide-gap plasma deposited a-Si:H-based alloys. Hydrogenated amorphous alloys of silicon and carbon (a-Si$_x$C$_{1-x}$:H$_y$) are an interesting variant to standard a-Si:H used in heterojunction solar cell devices. The addition of carbon adds extra freedom in order to control the properties of the material, as increasing concentrations of carbon in the alloy widen the optical bandgap in order to potentially increase the light efficiency of solar cells made with amorphous silicon carbide layers. The a-SiC:H alloys are fabricated by decomposition of silane (SiH$_4$), phosphine (PH$_3$), methane (CH$_4$) and hydrogen (H$_2$).Particularly, the investigation focused on the incorporation of hydrogen and carbon within the resulting [a-Si$_x$C$_{1-x}$(n):H$_y$] and [a-Si$_x$(n):H$_y$] films, which later form the emitter. The corresponding local vibrational modes of Si-H$_x$, C-H and corresponding network have been analyzed by μ-Raman spectroscopy. It is confirmed that the band gap E$_G$ can be tailored using an appropriate gas mixture during the decomposition. Furthermore the I-V characteristics of the prepared heterojunction solar cells are analyzed. A trade-off between electrical defects density and optical losses causes an improvement of the I-V characteristics with increasing carbon and hydrogen concentration in the feed stock during deposition. However, the problems due to carbon induced electronic defects, such as low conversion efficiency due to the sp^2/sp^3 bonding structure of C-C, as well as deterioration of the photo-conductivity, still should be resolved.

(iii) Investigations on wide-gap plasma deposited μc-Si:H-based alloys.
According to the film properties discussed, p^+ and n^+ μc-Si:H films are likely
to be suitable for use as emitter and BSF in a heterojunction solar cell device.
They indicate high transparency to suppress absorption, and high conductivity when annealed at the optimum temperature. By means of post-annealing
in a range of 180 °C < T_{ann} < 300 °C of the μc-Si:H(p) films, an activation of
the boron atoms occurs, which was formerly bound in form of B-H-Si. The
annealing step might release the hydrogen from this B-H-Si bound, leaving
activated B-Si bounds. Dark conductivities of σ_{dark} = 25 S/cm have been
achieved. For μc-Si:H(n) films, an annealing temperature in the range between
180 °C < T_{ann} < 250 °C is found to be an optimum temperature to improve
dark conductivity up to σ_{dark} = 130 S/cm.

**(iv) Heterojunction solar cell devices incorporating a-Si:H- and μc-Si:H-
based alloys.** Intrinsic a-SiO$_x$:H(i) films are formed in order to prove their
applicability for surface passivating buffer layers. In spite of the importance
of the a-SiO$_x$:H layer, almost no studies have focused on the structure and
properties of thin a-SiO$_x$:H layer incorporated in actual solar cell devices. Such
characterization is of significant importance to clarify the role of the a-SiO$_x$:H
layer in the heterojunction cell devices. The optimum film thickness is determined by a trade-off between passivation quality, current losses due to low
conductivity of the a-SiO$_x$:H films resulting in high series resistances, light
absorption and deposition homogeneity. By incorporating a-SiO$_x$:H(i) to the
heterojunction structure a drastic increase of the open circuit voltage (up to
655 mV for p-type substrates and 695 mV for n-type substrates) is found,
and accordingly, a higher conversion efficiency than obtained with standard
a-Si:H(i). These high open-circuit voltages can be consistently ascribed to the
adequate surface passivation by a-SiO$_x$:H preventing surface recombination
at the hetero-interface and to the decrease of the optical absorption in the
blue light region due to an enhanced optical bandgap of 1.95 eV. Heterojunction solar cells using textured substrates exhibited an expected J_{sc} gain, when
transferring the optimized process parameters of cells using polished sub-

strates to cells using textured substrates. For heterojunction solar cells using textured substrates, efficiencies exceeding 19 % have been obtained.

CONTENTS

CHAPTER 1.

INTRODUCTION

In one form or another, solar energy is the source of nearly all energy on earth. The solar photovoltaic energy conversion describes thereby a process generating electrical energy directly from sunlight (photon energy) using a solar cell device [1]. Photovoltaic (PV) is the as simple as elegant method of utilizing the photon energy by direct conversion of the incident solar radiation into electricity, with no noise, pollution or moving parts, making them robust, reliable and long lasting [2].

The photovoltaic effect was first discovered by *Becquerel* in *1839*. In 1900 *Max Planck* proposed that energy is not a continuous wave, as Newton thought, but occurred in discrete packets, called 'quanta'. Then in 1905 Einstein postulated that light consisted of these discrete packets (or quanta), later dubbed 'photons'. With this stroke of genius idea Einstein was able to explain the photoelectric effect, why electrons are emitted from metals driven by light illumination. Then in 1913 the Danish physicist Niels Bohr gave an entirely new picture of the atom, one that resembled a miniature solar system. But unlike in a solar system in outer space, electrons can only move in discrete orbits (shells) around the nucleus. When electrons 'jumped' from one shell to a lower shell with less energy, they emitted a photon of energy. When an electron absorbed a photon of a discrete energy, it 'jumped' to a higher shell with more energy. Today the photoelectric effect and the photon form the basis of solar cells (and much of modern electronics). [3]

Research and development of photovoltaic received its first major boost from the space industry in the 1960s. Solar cells became an interesting scientific variation to the rapidly expanding silicon transistor developments. The oil crisis in the 1970s focused the worlds attention on the desirability of alternate energy sources for terrestrial use, which in turn promoted the investigation of

photovoltaics as a promising alternative energy source. Although the oil crisis proved short-lived and the financial incentive to develop solar cells abated,
solar cells had entered the arena as a power generating technology. The advantage of solar cells as 'remote' power supply was quickly recognized and
prompted the development of a terrestrial photovoltaic industry. Small scale
transportable applications (such as calculators and watches) were utilized and
remote power applications began to benefit from photovoltaics [2].

In the 1980s, based on the achievements in research, solar cells began to
increase their efficiency. Over the next decade, the photovoltaic industry ex-
perienced steady growth rates of between 15 % and 20 %, largely promoted by
the remote power supply market. The year 1997 saw a growth rate of 38 %.
Today, solar cells are recognized not only as a method of power production
and improvement of life quality to those who do not have grid access, but
they became also a tool to significantly diminish the impact of environmental
damage caused by conventional electricity generation in advanced industrial
countries. Today, solar cells gathering global attention as increasingly important renewable alternative to conventional fossil fuel electricity generation. [2]

Figure 1.1 illustrates the best research cell efficiencies as a function of time
till today for essentially four classes of material: silicon-based, thin-film based,
multijunctions (including quantum dots at the research level), and emerging
organic materials. For single bandgap materials - like discussed in this thesis
- under 'one sun' conditions, the theoretical efficiency of 31 % is called the
Shockley-Queisser limit for quantum conversion.

Increasing the cell efficiency in industrial production has made great
progress in the last few years. Cell structures like the A300 of *SUNPOWER*
or the HIT™cell of *SANYO* are demonstrating the potential for industrial cells
to achieve efficiencies greater than 20 %. It is therefore, that this work is concerned about heterostructures and their optimization. It addresses some of
the most important cell components which need further improvement to reach
higher efficiencies [4]. The next section summarizes the main characteristics
of a heterojunction solar cell without discussing its physical function in detail.
The main characteristics of a heterojunction photovoltaic cell are discussed
in detail in section 2.3.4.

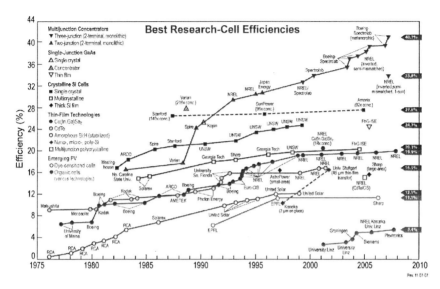

Figure 1.1.: Illustration of the best research cell efficiencies as a function of time for essentially four classes of material: silicon-based, thin-film based, multi-junctions (including quantum dots at the research level), and emerging organic materials. For single bandgap materials under 'one sun' conditions, the theoretical efficiency of 31 % is called the Shockley-Queisser limit for quantum conversion. The only cells with efficiencies above this are the multijunction cells, which are upper-bound by the thermodynamic limit of 68 %. (source: *NREL*).

1.1. Heterojunction solar cells: State of technology

Recently, a growing interest in a-Si:H/c-Si heterojunction solar cells is observed due to high solar cell efficiencies now exceeding 22 % [5]. Advantages of heterostructure based solar devices compared to a conventional c-Si homojunction device include suppression of efficiency reduction at high operating temperatures, and a rather simple and cost-effective solar cell processing that can be performed at low temperatures (< 200 °C). A considerable number of studies have been published describing a-Si:H/c-Si heterojunction solar cells (i.e., [5–17]). Nevertheless, there has been little consensus on the properties of the a-Si:H/c-Si hetero-interface, and thus a clear insight into the a-Si:H/c-Si solar cell has been lacking [18]. This is reflected in the discrepancy in solar cell performances of the various studies listed above.

First a-Si based heterojunction solar cells (a-SiC:H/a-Si:H/c-Si heterojunction), with a conversion efficiency of 9 %, were presented by *Hamakawa et al.* [19]. The concept of the heterojunction solar cell with intrinsic layer to suppress recombination was first introduced by *SANYO* [8]. The *p-n* junction was not fabricated by diffusion, but depositing a thin, doped layer of amorphous silicon on a crystalline silicon wafer by a plasma enhanced chemical vapor deposition (PECVD). By introducing an additional intrinsic, thin amorphous layer between the crystalline wafer and doped emitter, the HIT (Heterojunction with Intrinsic Thin Layer) cell concept was established. Today, high efficiency heterojunction solar cells reach up to 739 mV and 22.3 % efficiency on a approximatly 100 cm^2 area [5]. Also, they offer an alternative method of cost reduction: the more power a solar cell can generate, the fewer cells are necessary to obtain the same amount of power, reducing the amount of materials and total system costs needed [9]. The key-factor in enhancing solar energy conversion lies in achieving both high-efficiency and low-cost device production. However, to obtain such high efficiency cells, one have to concentrate on technological problems such as light trapping, surface passivation, improving grid electrodes for higher short-circuit current, reducing absorption losses, enhancing amorphous silicon properties for higher open-circuit voltage, et cetera. Recently, *von Roedern* [20] reported that solar cell optimization is ineffective, if optimization schemes focus on a single aspect of the device at a time. He suggested abandoning previously practiced solar cell optimization schemes of optimizing a single device feature at a time. *Wang et al.* [21] reported that the cell voltage of heterojunction solar cells improved when both the front and back contact are hetero-contacts. Conventional 'systematic' cell optimization studies that would change only one device aspect at a time (e.g., the front contact from a pin junction to a heterojunction contact) might actually miss the benefit of hetero-contacts by not implementing them simultaneously at the front and rear of the cell [20].

Similarly, in the nineteenth century, *Justus von Liebig* proposed his 'law of the minimum', which describes the yield potential of crop production as being limited by the nutrient in shortest supply. According to this law, crop

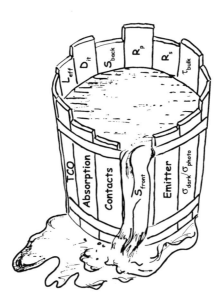

Figure 1.2.: Liebig's law adapted to solar cells. Since several loss mechanisms limit the cell potential at the same time, it is a simplified view of a complex task. After *Glunz* [4], slightly modified.

production is like a barrel with staves of unequal length. Today, one can adapt this law of the minimum in a simplified view of a complex task to describe the efficiency potential of a solar cell (illustrates in Fig. 1.2). The capacity of the barrel (i.e., the efficiency potential of the cell) is limited by the length of the shortest stave (in this case the surface recombination velocity at the front side, S_{front}). Of course this is a simplified view of a complex task, since several loss mechanisms can limit the cell potential at the same time [4].

1.2. Thesis motivation

The present thesis is concerned about optimization of heterojunction solar cells, where special attention is paid to (i) the front and back surface passivation of the bulk material (avoiding the protected device structure from *SANYO*) by high-quality wide-gap amorphous silicon sub-oxides (a-SiO$_x$:H), and (ii) the influence of wide-gap high-quality materials at the front side for use as emitter and passivation to suppress absorption losses, such as a-SiC:H, a-SiO$_x$:H, and μc-Si:H (standard a-Si:H(i) is applied as reference).

Wide-gap a-SiC:H, a-SiO$_x$:H, and μc-Si:H. The requirements of high optical depth and perfect charge collection imply very high demands of material quality. Hydrogenated amorphous silicon (a-Si$_x$:H$_y$) has been accepted as a suitable heterojunction material for a-Si:H/c-Si solar cells forming both the emitter and surface passivation layer, and conversion efficiencies exceeding 21 % have been reported recently [22]. Nevertheless, the light absorption in the a-Si:H(i) and the doped a-Si:H layers is rather strong, and the short-circuit current density in the a-Si:H/c-Si solar cells rapidly decreases with increasing a-Si:H layer thicknesses [18].

To improve the short circuit density in a-Si:H/c-Si solar cells further, it is preferable to employ an a-Si:H-based alloy that has larger optical bandgap than standard a-Si:H. Hydrogenated amorphous alloys of silicon and carbon (a-SiC:H), and silicon and oxygen (a-SiO$_x$:H) are an interesting variant to standard a-Si$_x$:H$_y$. The introduction of carbon, and oxygen with additional hydrogen adds extra freedom in controlling the optical and electrical properties of the material, as increasing concentrations of C, O, and H$_2$ in the alloy widen the electronic gap between conduction and valence bands. Therefore, one main topic of this thesis is the investigation of plasma enhanced chemical vapor deposited layers in order to prove the feasibility to widen the optical band gap in emitters of heterojunction solar cells. Particularly, this has been carried out by application of hydrogenated amorphous carbon-silicon alloys (a-Si$_x$C$_{1-x}$:H$_y$), amorphous silicon sub-oxides (a-SiO$_x$:H) and micro-crystalline layers (μc-Si:H), each compared to standard a-Si$_x$:H$_y$ layers.

The fraction of incident light absorbed in the front layer does not contribute to the photocurrent. This limitation can be removed if the front layer is replaced by a wide bandgap semiconductor since such a layer will be relatively transparent [23]. The photovoltaic performance of the heterojunction device depends on the band structure of both semiconductors (*p-n* junction), the position of their Fermi levels and the nature of any interface states. Figure 1.3

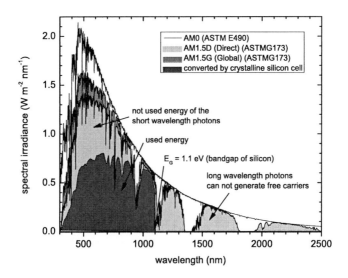

Figure 1.3.: Standard solar spectra. The standard spectrum for space applications is referred to as AM0. It has an integrated power of 1366.1 W/m². Two standards are defined for terrestrial use. The AM1.5 Global spectrum is designed for flat plate modules and has an integrated power of 1000 W/m² (100 mW/cm²). The AM1.5 Direct spectrum is defined for solar concentrator work. After [2].

displays the standard reference spectra, illustrating the wavelength range used for c-Si/a-Si:H hetero-structures.

Surface recombination. As wafers are thinned – driven by the need to conserve Si material – the passivation of front and back surfaces becomes more critical [20]. In heterojunction solar cells, the a-Si:H/c-Si hetero-interface is of significant importance, since the hetero-interface characteristics directly affect the junction properties and thus solar cell efficiency. Also, the temperature dependence of the solar cell performance has been recognized as one of the most important characteristics for determing output power in practical use. Well surface-passivated solar cells, in other words, high-V_{oc} cells tend to exhibit better temperature dependencies compared to traditional c-Si solar cells (*cf.* [24, 25]) [9].

The prime importance therefore is to passivate the c-Si surfaces properly. Passivation schemes commonly used in photovoltaic applications are silicon dioxide (SiO_2) [26], silicon nitride (SiN_x) [27, 28], but also intrinsic amorphous silicon (a-Si:H(i)) [6], amorphous silicon carbide [29] and stacks of those. The insulating SiO_2 and SiN_x for use as passivation in heterojunction devices can be neglected. The use of a-SiC:H is inferior compared to a-Si:H(i), *cf.* [29]. Hydrogenated amorphous silicon (a-Si:H(i)) has been accepted as a suitable heterojunction material for a-Si:H/c-Si solar cells, but thus protected in its use by *SANYO* for commercially available heterojunction cells. A novel passivation scheme, amorphous silicon sub-oxides (a-SiO_x:H) is introduced and investigated intensively in this work.

1.3. Outline

Chapter 2   starts with a brief discussion of basic aspects of heterojunction solar cell physics. In particular, it is concerned with the mechanisms of charge carrier generation, recombination theory, and different recombination paths and the influence of those on the solar cell efficiency. In prospect of a-Si:H/c-Si heterojunction solar cells investigated in this work, a brief discussion about these types of cells will be given, followed by optical and electrical properties which will be described within the context of the electronic band theory of solids, and the nature of the amorphous material used for fabrication.

Chapter 3   presents on overview on the basic steps of sample preparation, the deposition of amorphous silicon by PECVD, and describes fundamental analysis techniques used in this work. Special attention is paid to the applied characterization methods to investigate the optical, electrical and morphological properties of plasma enhanced deposited a-Si:H/μc-Si:H layers. These are (i) the determination of the bandgap of a semiconductor by use of spectroscopic ellipsometry, (ii) the corresponding local vibrational modes analysis of amorphous and micro-crystalline layers by use of μ-Raman spectroscopy,

(iii) cell characterization methods in respect of heterojunction devices, as well as (iv) lifetime measurements using either quasi-steady-state-, transient- or microwave-photoconductance decay are described.

Chapter 4 contains an investigation on the incorporation efficiency of hydrogen and carbon in a-Si_xC_{1-x}:H_y and a-Si_x:H_y films. A systematic analysis of theirs influence on electrical and optical properties is given. It is shown that beside the ability to change the electronic properties of hydrogenated amorphous silicon, the addition of carbon enhances the optical band gap and suppresses absorption in the thin film emitter. Heterojunction solar cells prepared with wide-gap a-Si_xC_{1-x}:H_y and wide-gap a-Si_x:H_y as emitter layer are compared.

Chapter 5 reports on PECV deposited micro-crystalline layers or close to the microcrystallinity grown amorphous films. Both p-type and n-type films are fabricated using PH_3 and TMB, respectively. The role of the PECV deposition parameters is investigated for optimizing the μc-Si:H films in terms of dark conductivity. In addition, electrical and optical properties are systematically studied using various analytical methods, such as μ-Raman spectroscopy and spectroscopic ellipsometry. The μc-Si:H layers feature a high optical bandgap and high electrical conductivity. The impact of post-annealing of the μc-Si:H films is investigated in terms of electrical conductivity and change in the micromorph network.

Chapter 6 presents a systematic study of the surface recombination properties of PECV deposited amorphous silicon sub-oxides (a-SiO_x:H). Those a-SiO_x:H films are fabricated by decomposition of silane, hydrogen and carbon-dioxide. The effects of the PECV deposition parameters on the effective lifetime and hence on the surface recombination velocity are investigated. Outstanding high effective lifetimes are detected exceeding those of record SiO_2 and SiN_x passivated silicon wafers of similar doping type and resistivity. A compositional analysis of the a-SiO_x:H is presented, where μ-Raman spectroscopy, spectroscopic ellipsometry, quasi-steady-state lifetime measurements, transmission and absorption measurements and secondary ion-mass spectroscopy are combined to investigate those a-SiO$_x$:H films in detail. It is demonstrated that apart from their excellent surface passivation quality, they also show less absorption in the blue-light region.

Chapter 7 reports on experimental heterojunction silicon solar cells passivated with wide-gap a-SiO$_x$:H and wide-gap μc-Si:H emitter and back-surface-field, which are developed in the previous chapters. A variety of cell designs have been compared, incorporating both n-type and p-type float-zone silicon wafers. With the use of a-SiO$_x$:H, very high open circuit voltages up to 700 mV are demonstrated.

Chapter 8 summarizes the research of this thesis and some remarks on future work is outlined. The investigations presented in this thesis are based mainly on the publications given in chapter 'list of publications'.

Chapter A contains a brief description of general process steps for the fabrication of heterojunction solar cells such as cleaning of the bulk material, texturing, fabrication of an anti-reflection-coating and formation of metal contacts. In addition a list of available gaseous precursors for the PECV deposition of a-Si:H and its counterparts is given.

CHAPTER 2.

BASIC ASPECTS OF HETEROJUNCTION SOLAR CELL PHYSICS

The most important parameters of a semiconductor material for solar cell operation are the band gap, the number of free carriers available for conduction, and the 'generation' and 'recombination' of free carriers in response to incident photons. Therefore, the following pages cover the characteristics of a solar cell, basic aspects of semiconductor materials, and the corresponding physical mechanisms in matters of heterojunction solar cells. However, this chapter is not starting with the assumption that the reader has no familiarity with the subjects whatsoever; more detailed physics can be found elsewhere (i.e., [1, 24, 30–34]).

2.1. Generation, recombination, and carrier lifetime in silicon

This section is concerned about charge carrier generation, different recombination paths and the influence of those on solar cell efficiency. As for a solar cell device, the essential characteristic is the generation of a photocurrent. The output of this device is determined by light absorption, current generation and charge recombination [31].

2.1.1. Generation

The generation of electrons and holes in semiconductors may occur due to light absorption, if the photon energy, E_{ph}, is large enough to raise an electron from the valence band into an empty conduction band state. Carrier generation may also occur due to the transition of an electron from the valence band into a localized state in the bandgap, which generates only a hole, or from a localized state into the conduction band, which generates only an electron. However, generation is an electronic excitation process, which requires an input of energy (while recombination releases energy). This energy can be provided not only by incident light (photons) but also by the vibrational energy of the lattice (phonons) or the kinetic energy of another carrier.

In the particular case of photovoltaic devices, photogeneration is by far the most important process; other significant generation processes are trap-assisted generation or Auger generation. If an incident photon is absorbed in the solar cell device, it creates both a majority and a minority carrier. In photovoltaic devices, the number of light-generated carriers are a few orders of magnitude less than the number of majority carriers already present in the solar cell due to doping. Consequently, the number of majority carriers in an illuminated semiconductor does not alter significantly. However, the opposite is true for the number of minority carriers. The number of photo-generated minority carriers is larger than the number of minority carriers existing in the solar cell in the dark, and therefore the number of minority carriers in an illuminated solar cell can be approximated by the number of light generated carriers [2].

Under steady-state-conditions, a generalized form of three basic equations can be used to address the relationship between carrier generation, recombination and transport within silicon. These basic equations are *Poissons* equations and the carrier continuity equations, which are discussed in more detail elsewhere (*cf. van Roosbroeck* [35], *Sze* [36]). The *Poissons* equation can be expressed as

$$\nabla \cdot \epsilon \vec{E} = q \left(p - n + N \right),$$ (2.1)

where N is the net charge due to dopants and other trapped charges, which can be written as $N = (N_D^+ - N_A^- + N_T^+)$, $\vec{E}$ is the electric field vector, and q is the electronic charge. Hole and electron continuity equations are

$$\nabla \cdot \vec{J_p} = q \left(G - R_p - \frac{\partial p}{\partial t} \right) \quad \text{and} \quad \nabla \cdot \vec{J_n} = q \left(R_n - G + \frac{\partial n}{\partial t} \right),$$ (2.2)

where G is the optical generation rate of electron-hole pairs. Thermal generation is included in R_p and R_n. J_n and J_p are the electron and hole current density vectors.

A key factor in determining if a photon is absorbed or transmitted is the energy of the incident photon. Photons incident onto a crystalline wafer material can be divided into three groups based on their energy compared to that of the semiconductor bandgap [2]:

1. $E_{ph} < E_G$

   Photons with the energy E_{ph} less than the bandgap energy E_G interact only weakly with the semiconductor, passing through it as if it would be transparent.

2. $E_{ph} = E_G$

   Photons with the energy E_{ph} equal to the bandgap energy E_G have just enough energy to create an electron hole pair and are efficiently absorbed.

3. $E_{ph} > E_G$

   Photons with the energy much greater than the bandgap are effectively absorbed.

The generation rate G for a given photon energy can be expressed via the photon flux, N_{ph}. The photon flux is attenuated proportional to the number of photons which are annihilated in the absorption process

$$G = \alpha_{eh}N_{ph} = \alpha\frac{qP_{opt}}{E_{ph}}, \tag{2.3}$$

where α_{eh} is a component of the absorption coefficient (α) corresponding to the electron-hole pair creation. In case a distribution of photons of different wavelengths is involved, N_{ph} becomes a spectral distributed photon flux and one has to integrate over the wavelength to obtain G as

$$G = \int_{\lambda_{min}}^{\lambda_{max}} \alpha_{eh}(\lambda)N_{ph}(\lambda)d\lambda. \tag{2.4}$$

2.1.2. Bulk recombination

The loss of mobile electrons or holes by any kind of removal mechanisms can be referred to a recombination process. In general, the recombination of electrons and holes is a process at which both carriers annihilate each other, so to say electrons occupy (through one or multiple steps) the empty state associated with a hole. Both carriers eventually disappear in the process. The energy difference between the initial and final state of the electron is released in that process. This leads to one possible classification of the recombination processes. The energy released can be given up as a photon (*radiative recombination*), as heat through phonon emission (*non-radiative emission*) or as kinetic energy to another free carrier (*Auger recombination*).

For a photovoltaic device, several different recombination mechanisms of electrons and holes in a semiconductor are important. Those mechanisms can be divided into two groups: *intrinsic* (unavoidable) recombination processes, occuring due to essential physical processes, and *extrinsic* (avoidable) recombination processes, which can in principle be overcome by careful preparation of the material (*cf*. *Haecker and Hangleiter* [37]). The intrinsic recombination mechanisms are band-to-band radiative recombination, which results from optical generation, spontaneous emission, and *Auger* recombination. These processes describe the interaction of an electron or hole with a second similar carrier, resulting in decay and increase of kinetic energy each with one of the carriers. The extrinsic recombination processes usually involve recombination through deep impurities or defects in the bandgap (trap states), due to imperfect material.

Other major recombination mechanisms within a photovoltaic device can be generally described with the following fundamental processes: the high number of dangling bonds at the surface of a crystalline wafer creates a large density of defects throughout the bandgap. Surface recombination is therefore a particular case of the recombination methods above. Likewise, emitter recombination can be described as a particular case of the recombination methods, as in the heavily doped emitter region of a photovoltaic device the high doping results in Auger recombination, limiting the lifetime in the emitter. [38]

The recombination methods and processes mentioned above will be described in detail in the following to understand and quantify effective lifetime measurements (which are carried out later in this thesis) in terms of device characteristics. Commonly the carrier lifetime τ can be defined as

$$\tau\left(\Delta n, n_0, p_0\right) = \frac{\Delta n}{U\left(\Delta n, n_0, p_0\right)}, \tag{2.5}$$

where U defines the net recombination rate. As can be seen, τ depends significantly on the injection conditions Δn (except in the case U ~ Δn), as well as on the dopant level of the material through n_0 ($\Delta n \equiv n - n_0$) and p_0 ($\Delta p \equiv p - p_0$). Detailed theory of recombination in semiconductors are well covered by e.g. *Landsberg* [39], *Shur* [40] or *Tyagi* [41].

Analog to the definition of the recombination lifetime (Eq. 2.5), the *surface recombination velocity (SRV)*, S, at the semiconductor surface is defined as

$$U_S = S\Delta n_S \tag{2.6}$$

where Δn_S is the excess minority carrier density at the surface. The SRV is typically used to quantify the surface recombination processes, and hereafter used along with the measured effective lifetime, τ.

2.1.3. Radiative recombination

Radiative recombination constitutes the reverse process of photon absorption in a semiconductor; it is simply speaking a spontaneous emission process, as depicted in Fig. 2.1. In a spontaneous emission process, an electron is falling from an allowed conduction band into a vacant valence band state. The energy difference between the initial and final state of the electron is released in the process. In the case of radiative recombination, this excess energy is mainly emitted in the form of a photon. The energy of the emitted photon usually corresponds to an energy close to that of the bandgap.

The radiative recombination can be characterized by the radiative lifetime $\tau_{radiative}$, which depends on the doping type and concentration in a wafer, as

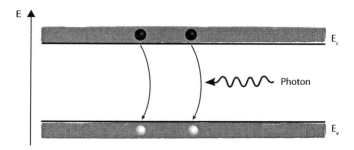

Figure 2.1.: Energy band visualization of radiative recombination. [34]

well as the injection level (Δn) or excess carrier density (ECD), and is given by

$$\tau_{rad} = \frac{1}{B\left(n_0 + p_0 + \Delta n\right)},\qquad(2.7)$$

where B is a material constant; for silicon at 300 K it has a relatively small value of $1 \cdot 10^{-14}$ cm^{-3}/s due to nature of silicon as an indirect semiconductor. Separated for high- and low-injection it can be expressed via

$$\tau_{rad,li} = \frac{1}{BN_{dop}} \qquad \text{and} \qquad \tau_{rad,hi} = \frac{1}{B\Delta n},\qquad(2.8)$$

whereas $\tau_{rad,li}$ corresponds to the radiative lifetime under low injection and $\tau_{rad,hi}$ analog to high injection level; N_{dop} is the density of donor (N_D) or acceptor (N_A) atoms, respectively [38]. As seen from equation 2.8, at low injection conditions the radiative lifetime does not depend on the injection level: the radiative recombination is almost always much larger than the Auger lifetime and produces negligible recombination rates in Si compared to Auger recombination and recombination through defects and impurities. Therefore, the radiative lifetime can be neglected. The impact of radiative recombination on the carrier lifetime is depicted in section 3.6.

2.1.4. Auger recombination

The Auger recombination can be described as a three particle interaction: a collision between two similar carriers results in the excitation of one carrier to

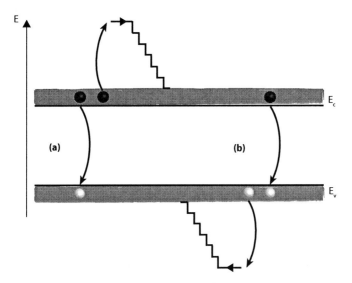

Figure 2.2.: Schematic energy-band visualization of Auger recombination. The excess energy is transferred to: **(a)** a conduction band electron or **(b)** a valence band hole. The excess energy dissipates as heat as the excited carrier relaxes to the band edge. [38]

a higher kinetic energy, and the recombination of the other across the bandgap with a carrier of opposite polarity. A schematic of this interaction is depicted in Fig. 2.2, after [31, 42]. The energy released through recombination is dissipated to the other carrier, which is either an electron in the conduction band or a hole in the valence band. Either an electron and two holes (with a recombination rate $U_{ehh} = C_p np^2$) or a hole and two electrons (with a recombination rate $U_{eeh} = C_n n^2 p$) are involved in band-to-band Auger recombination. In total, the net Auger recombination rate, U_{Auger}, is given by the sum of U_{ehh} and U_{eeh}

$$U_{Auger} = C_n \left(n^2 p - n_0^2 p_0 \right) + C_p \left(np^2 - n_0 p_0^2 \right), \tag{2.9}$$

where C_n and C_p are called Auger coefficients. The common Auger recombination for n-type doped material and p-type doped material of the electron lifetime can be characterized by the Auger lifetime τ_{Auger}, which depends only on the doping type and concentration in a wafer, as well as on the injection level (excess carrier concentration) Δn. The Auger lifetime is given by

$$\tau_{Auger} = \frac{1}{c_1 + c_2 \Delta n + c_3 \Delta n^2},\tag{2.10}$$

where $c_1 = C_n n_0^2 + C_p p_0^2$, $c_2 = C_n n_0 + C_p p_0 + (C_n + C_p)(n_0 + p_0)$, and $c_3 = C_n + C_p$. Under low-injection ($\tau_{Auger,li}$) is given for

n-type doped material $\qquad\qquad \tau_{Auger,li} = \dfrac{1}{C_n N_D^2}$ $\qquad\qquad$ (2.11)

p-type doped material $\qquad\qquad \tau_{Auger,li} = \dfrac{1}{C_p N_A^2}$ $\qquad\qquad$ (2.12)

where $C_a \equiv C_n + C_p$ is the ambipolar Auger coefficient. To determine C_n and C_p experimentally, the carrier lifetime in the bulk of a c-Si material is measured under low-injection and the corresponding Auger coefficient is extrapolated from the fit of τ_{Auger} vs. N_{dop}. Under high injection conditions, $\tau_{Auger,hi}$ becomes independent on the doping type and can be expressed for both doping types as

$$\tau_{Auger,hi} = \frac{1}{\left(C_n + C_p\right)\Delta p^2},\tag{2.13}$$

Measured values for Auger coefficients are in the range of $C_n = (1.7\text{-}2.8) \cdot 10^{-31}$ cm^{6}s^{-1} and $C_p = (0.99\text{-}1.2) \cdot 10^{-31}$ cm^{6}s^{-1} for c-Si, which proved to be constant at dopant concentrations greater than $5 \cdot 10^{18}$ cm^{-3}, after *Dziewior and Schmid* [43] and *Beck and Conradt* [44]. However, for dopant concentrations smaller than $5 \cdot 10^{18}$ cm^{-3}, smaller lifetimes are measured experimentally than predicted with above Auger coefficients. *Haecker and Hangleiter* [37] concluded that the reason for the lower Auger coefficients measured are most likely Coulomb-enhanced interactions of the charge carriers. This Coulomb-enhanced Auger recombination is approximated to be

$$\tau_{Coulomb-Auger} = \frac{2.374 \cdot 10^{24}}{N_{dop}^{1.67}} s,\tag{2.14}$$

valid for low injected p-type c-Si. More details on this theory can be found in [28, 45, 46]. The impact of Auger-recombination on the carrier lifetime is depicted in section 3.6.

2.1.5. Trap assisted recombination through impurities and defects

Besides the possibility of intrinsic recombination discussed in previous section, extrinsic recombination occurs when an electron falls into a trap, an energy level within the forbidden bandgap of a semiconductor caused by the presence of a foreign atom (impurity) or crystallographic imperfections such as dislocations. These impurities or defects can be found in almost any semiconductor and are responsible for a 2-step recombination process. First analyzed by *Shockley and Read-Jr.* [47] and independently by *Hall* [48] in 1952, the Shockley-Read-Hall (SRH) model considers the carrier recombination through defect levels with a fixed thermal energy, E_t. The model describes the path of an electron transition from the conduction band into a defect level at E_t and from there into the valence band, subsequent recombining with a hole. The excess energy released during the recombination event is dissipated by lattice vibrations or phonons. Figure 2.3 exhibits the energy band visualization of four possible interactions of electrons and holes with a defect level at E_t: (1) an empty defect level captures an electron; (2) a filled defect level emits an electron; (3) a filled defect level captures a hole; and (4) an empty defect level emits a hole. As deduced from Fig. 2.3, the recombination process can be divided into a two-step transition: once an electron occupying the trap, the trap can not accept another electron, and the trapped electron moves in a second step into an empty valence band state, thereby completing the recombination process.

Shockley and Read-Jr. [47] and *Hall* [48] made an attempt in theirs model to calculate the recombination rate U_{SRH} (in unit cm^{-3} s^{-1}) only by the recombination through one defect with fixed energy E_t. This defect may capture carriers with characteristic capture cross sections, σ_p and σ_n for the specific defect. The recombination rate, U_{SRH}, for a single defect level can be then

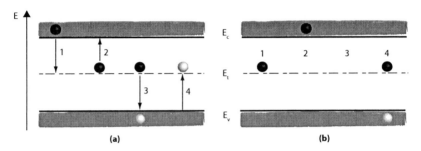

Figure 2.3.: Energy band visualization of recombination through a defect state within the energy bandgap. Four possible interactions of electrons and holes with a defect level at E_t are described *before* (Fig. 2.3a), indicated by the arrows, and *after* (Fig. 2.3b) the electron transition. After *Aberle* [45].

evaluated as

$$U_{SRH} = \frac{v_{th} N_t \left(np - n_i^2 \right)}{\frac{n+n_1}{\sigma_p} + \frac{p+p_1}{\sigma_n}} = \frac{np - n_i^2}{\tau_{n0} \left(p + p_1 \right) + \tau_{p0} \left(n + n_1 \right)}, \qquad (2.15)$$

where v_{th} is the thermal velocity of the charge carriers ($\sim 10^7$ cm/s in Si at 300 K), and N_t is the density of recombination defects. τ_{p0} and τ_{n0} state the fundamental electron and hole lifetimes related to the thermal velocity of charge carriers given by

$$\tau_{p0} = \frac{1}{\sigma_p v_{th} N_t} \qquad \text{and} \qquad \tau_{n0} = \frac{1}{\sigma_n v_{th} N_t}. \qquad (2.16)$$

The densities n_1 and p_1 express the case, when the Fermi-level and the defect-level are consistent. Carrier densities in conduction- and valence band are given by:

$$n_1 = N_C exp \left(\frac{E_t - E_C}{kT} \right) \quad \text{and} \quad p_1 = N_V exp \left(\frac{E_C - E_G - E_t}{kT} \right), \qquad (2.17)$$

where N_C and N_V are the effective density of states at the conduction and valence band edges, E_t is the energy level of the defect and E_C and E_G are the conduction band and bandgap energies.

Using $U \equiv \Delta n / \tau_{SRH}$ and following Eq. 2.15, the SRH lifetime can be deduced as

$$\tau_{SRH}(\Delta n) = \frac{\tau p_0 (n_0 + n_1 + \Delta n) + \tau n_0 (p_0 + p_1 + \Delta p)}{p_0 + n_0 + \Delta n} \tag{2.18}$$

As can be seen, the SRH lifetime in Eq. 2.18 depends on the excess carrier lifetime. For low-injection conditions, Δn is much smaller than the doping concentration of the silicon wafer; hence, Δn can be neglected for both doping types. The SRH lifetime for low-injection conditions becomes therefore

$$n\text{-type doped material} \quad \tau_{SRH,li,n} = \tau_{n0} \left(\frac{p_1}{n_0} \right) + \tau_{p0} \left(1 + \frac{n_1}{n_0} \right) \tag{2.19}$$

$$p\text{-type doped material} \quad \tau_{SRH,li,p} = \tau_{p0} \left(\frac{n_1}{p_0} \right) + \tau_{n0} \left(1 + \frac{p_1}{p_0} \right) \tag{2.20}$$

The most effective recombination centers are deep defects, featuring energies close to the bandgap. For those, n_1 and p_1 are significantly lower than the dopant level of the silicon wafer (in case the defect level is exactly in the middle of the bandgap, $n_1 = p_1 = n_i$). For such deep defects, the SRH lifetime for low injection conditions can be simplified to

$$\text{for } n\text{-type material} \quad \tau_{SRH,li,n} = \tau_{p0} \tag{2.21}$$

$$\text{for } p\text{-type material} \quad \tau_{SRH,li,p} = \tau_{n0} \tag{2.22}$$

For high-injection conditions, Eq. 2.18 can be simplified to

$$\tau_{SRH,hi} = \tau_{n0} + \tau_{p0}. \tag{2.23}$$

The impact of trap assisted recombination (SRH) on the carrier lifetime is depicted in section 3.6.

2.1.6. Surface recombination through impurities and defects

Surface recombination determines the impact on both the short-circuit current (J_{sc}) and the open-circuit voltage (V_{oc}). Any defect or impurity within or at the surface of a silicon wafer promotes a recombination of carriers. The surface or interface of crystalline silicon represents thereby a severe discontinuity of the crystal lattice. This discontinuity of the crystal structure leads to a formation of 'dangling bonds', whose energy levels are allocated partly within in the bandgap near the semiconductor surface and are therefore effective recombination centers, causing a high local recombination rate as illustrated in Fig. 2.4. This high recombination rate at the surface depletes the region of minority carriers. In turn, a localized region of low carrier concentration causes a flow of carriers into this region from the surrounding; therefore, the surface recombination rate is limited by the rate at which minority carriers move towards the surface [2].

High recombination rates at the 'top' surface of a silicon solar cell have a particularly detrimental impact on the J_{sc} since the top surface also corresponds to the highest generation region of carriers in the solar cell. Therefore, it is essential to minimize the density of surface states, in particular in respect of the high surface/volume ratio for solar cells. Minimizing the density of surface states is typically accomplished by reducing the number of dangling bonds, and hence the high surface recombination: by growing a layer on top of the semiconductor surface which ties up some of these dangling bonds, a *surface passivation* is obtained. The *surface recombination velocity* determines the recombination at a surface. At a surface with no recombination, the movement of carriers towards the surface is zero, and hence the surface recombination velocity is zero. In case a surface exhibits infinitely fast recombination, the movement of carriers towards this surface is limited by the maximum velocity they might attain [2].

2.1.6.1. Recombination through surface states

To determine the rate of surface recombination, U_S, for a single defect at the surface, the SRH analysis of section 2.1.5 can be applied:

$$U_S = \frac{n_s p_s - n_i^2}{\frac{n_s + n_1}{S_{p0}} + \frac{p_s + p_1}{S_{n0}}} \tag{2.24}$$

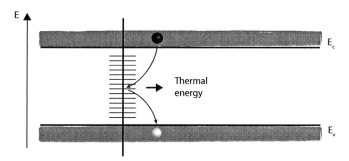

Figure 2.4.: Energy band visualization of recombination at a semiconductor surface, where multiple defect states exists. [34]

whereas S_{n0} and S_{p0} represents the surface recombination velocity parameters of electrons and holes, respectively, ($S_{n0} \equiv \sigma_n v_{th} N_{ts}$ and $S_{p0} \equiv \sigma_p v_{th} N_{ts}$, with N_{ts} describing the density of recombination defects at the surface), and n_s and p_s are the concentrations of electrons and holes at the surface. As can be seen from Eq. 2.24, the used SRH formalism has been reformulated in terms of recombination events per unit surface area, rather than per volume [38]. The main difference between recombination through defects at the surface and recombination through defects in the volume (*cf.* section 2.1.5) is the energetic distribution within the bandgap: whereas volume states exhibit discrete energy levels, the defect levels of the interface can be considered to be continuously distributed throughout the bandgap, caused (in case of dangling bonds) by static fluctuations of the bonding angle and the distances to next neighbor atoms of the defects. Both, their density and capture cross sections of the defect levels at the surface will be dependent on their energy level. By integration of Eq. 2.24, the surface recombination rate can be defined as

$$U_S = \left(n_s p_s - n_i^2\right) \int_{E_V}^{E_C} \frac{v_{th}}{\frac{n_s + n_1}{\sigma_p(E)} + \frac{p_s + p_1}{\sigma_n(E)}} D_{it}(E) dE \qquad (2.25)$$

whereas n_s and p_s are the electron and hole densities at the surface and D_{it} describes the density of states (states per unit surface area and energy). As the surface recombination rate U_S represents a rate per surface area, there is no point to define a surface lifetime according to Eq. 2.5. Instead, in analogy to Eq. 2.5 a *surface recombination velocity* (SRV), S, can be defined as

$$S\left(\Delta n_s, n_0, p_0\right) = \frac{U_s\left(\Delta n_s, n_0, p_0\right)}{\Delta n_s}, \tag{2.26}$$

whereas $\Delta n_s = n_s\text{-}n_{s0} = p_s\text{-}p_{s0}$ describes the excess carrier density at the surface. Considering equations 2.24 and 2.26, it becomes clear that the low- and high-injection level dependence of S arising from Eq. 2.26 can not be calculated as simple as in case of SRH-recombination (*cf.* section 2.1.5).

$$S\left(\Delta n_s, n_0, p_0\right) = \left(n_0 + p_0 + \Delta n_s\right) \int_{E_V}^{E_C} \frac{v_{th}}{\frac{n_0 + n_1 + \Delta n_s}{\sigma_p(E)} + \frac{p_0 + p_1 + \Delta n_s}{\sigma_n(E)}} D_{it}(E) dE \tag{2.27}$$

A large number of energy dependent variables approves the SRV to decrease, increase or remain constant as the injection level increases. However, the injection level dependence of S has been discussed in detail by *Aberle* [45] and *Schmidt* [28]. According to theirs work, a simple correlation for low injection can be deduced from Eq. 2.27 only for a maximum value of the SRV:

$$\text{n-type doped material} \qquad S_{li} = v_{th} D_{it} E_G \sigma_n \tag{2.28}$$

$$\text{p-type doped material} \qquad S_{li} = v_{th} D_{it} E_G \sigma_p \tag{2.29}$$

For high injection level, a similar correlation can be deduced for a maximum value of SRV, independently from the doping type [28]:

$$\text{n-type and p-type doped material} \qquad S_{hi} = \frac{v_{th} D_{it} E_G \sigma_n}{1 + \frac{\sigma_n}{\sigma_p}} \tag{2.30}$$

Following the results presented by *Luke and Cheng* [49], a general expression for the SRV in transient measurements can be expressed as

$$S = \sqrt{D\left(\frac{1}{\tau_{eff}} - \frac{1}{\tau_{bulk}}\right)} tan\left[\frac{W}{2}\sqrt{D\left(\frac{1}{\tau_{eff}} - \frac{1}{\tau_{bulk}}\right)}\right]. \tag{2.31}$$

From Eq. 2.31, approximate solutions for several cases can be found as follows:

- In case the surfaces are identical treated, $S = S_1 = S_2$

$$\tau_s = \frac{W}{2S} + \frac{1}{D}\left(\frac{W}{\pi}\right)^2 . \tag{2.32}$$

- In case one surface is perfectly passivated, so $S_2 = 0$

$$\tau_s = \frac{W}{S_1} + \frac{4}{D}\left(\frac{W}{\pi}\right)^2 . \tag{2.33}$$

- In case both surfaces are perfectly passivated, so $S_1 = S_2 = 0$

$$\tau_s = \infty. \tag{2.34}$$

- In case both surfaces have high recombination, so S_1 and S_2 are large

$$\tau_s = \frac{1}{D}\left(\frac{W}{\pi}\right)^2 . \tag{2.35}$$

- In case one surface has a high recombination and the other has a low recombination, so $S_1 = 0$ and $S_2 = \infty$

$$\tau_s = \frac{4}{D}\left(\frac{W}{\pi}\right)^2 . \tag{2.36}$$

2.1.6.2. Reducing surface recombination losses

Methods used to reduce the surface recombination rate are based on the following considerations:

Reduction of density of surface states From equation 2.27 can be concluded that reducing the density of interface states D_{it} results in a decrease of SRV. Hence, the properties of the surface states have to be optimized. To reach that, e.g. growing an appropriate dielectric layer such as thermally grown silicon oxide layer (SiO_2) is widely used in micro technology as well as for high efficiency solar cells: the silicon dioxide grows into the silicon, resulting in a new Si/SiO_2 interface with significant lower surface state density. Therefore, many of the dangling bonds are passivated by oxygen and hydrogen atoms and the density of interface states, D_{it}, is reduced [50]. In general, already a SiO_2 film of 10-15 Å thickness arise from the storage of a silicon substrate at air-conditions, but this film does not exhibit any surface passivation qualities. The preparation of an absolute clean Si-surface is experimentally complex. In this work investigated Si-surfaces and interfaces will be passivated with alternatively passivation schemes, such as plasma enhanced grown amorphous silicon oxides.

Reduction of concentration of free carriers at the surface   From Eq. 2.25 can be concluded that the recombination rate U_s depends on the excess concentration of the minority carriers. As the SRH recombination involves one electron and one hole, the recombination rate U_s reaches its maximum for the case both the surface concentrations n_s and p_s are approximately equal. Therefore, reducing the electron (n_s) or hole (p_s) density at the silicon surface will reduce the recombination rate. Technologically such a reduction of the surface concentration of one carrier type (either electrons or holes) can be realized by the formation of an electric field (built-in field) below the semiconductor surface:

- Application of a doping profile to the silicon wafer surface creates a high-low-junction (p^+-p or n^+-n), which decrease the minority carrier density at the surface. This structure – if applied to a solar cell – can appear i.e. as back surface field (BSF) (*cf.* [51]) or front surface field (FSF). It can be characterized by a saturation density J_0.

- Field effect passivation, which occurs due to an electronic field near the surface, essentially established by fixed charges in an overlaying dielectric layer, such as thermal grown silicon oxide or plasma deposited amorphous silicon and silicon nitride. The electric field enhanced by the fixed charges in the dielectric layer reduces the recombination rate by either repelling the minority carriers such as in an emitter region or a BSF (for *n*-type silicon material positive charges would repel the free electrons), or in an extreme case invert the surface (for *n*-type silicon material, large amounts of fixed negative charges would invert the surface), *cf.* [28, 38, 45].

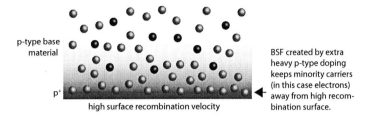

Figure 2.5.: Impact of an applied back-surface-field on the minority carriers. Heavy doping at the rear of the cell keeps the minority carriers away from high recombination rear contact. After [2].

2.1.6.3. Impact of the rear surface

To minimize the impact of rear surface recombination velocity on voltage and current in the case that the rear surface is closer than the diffusion length to the junction, a higher doped region at the rear surface of the solar cell *'back surface field'* (BSF) is used in general, as depicted in Fig. 2.5. The interface between the high and low doped region behaves like a p-n junction, and an electric field forms at the interface which introduces a barrier to minority carrier flow to the rear surface [2]. Since the minority carrier concentration is maintained at higher levels in the undoped region, the BSF has a net effect of passivating the rear surface.

2.1.7. Recombination in the emitter

The recombination in the emitter and in high-low-junctions (such as back-surface-fields (BSF)) can be characterized by a saturation current density, J_0, which represents actually the combined effects of Auger, SRH and surface recombination in the highly doped region. The emitter recombination can be regarded as a simplifying case of surface recombination: since the excess carrier concentration in the emitter is less than in the bulk silicon wafer, and the emitter is much thinner than the bulk, the emitter contribution to the photoconductivity is negligible. In that case, a recombination current into the emitter, J_{rec} can be defined after *Kane and Swanson* [52] as

$$J_{rec} = \frac{np}{n_i^2} J_{0E},$$ (2.37)

whereas J_{0E} is defined as emitter saturation current density. The recombination rate, U_E can be then expressed as:

$$U_E = \frac{J_{rec}}{qW} = \frac{np}{qWn_i^2} J_{0E},$$ (2.38)

whereas q describes the elementary charge of an electron, W is the width of the base region, and n and p refer to the electron and hole densities on the base side of the space-charge region. In high level injection, the emitter recombination rate is proportional to the square of the carrier concentration, hence it can be separated from the linear bulk and surface recombination [52].

For sample structures with highly doped emitters and high-low-junctions such as p^+nn^+ or p^+np^+, the lifetime at low and high injection level for *p*- and *n*-type base material can be defined as

$$\tau_{E,BSF,li} = \frac{qn_i^2 W}{J_{0E}N_{dop}} + \frac{qn_i^2 W}{J_{0BSF}N_{dop}}$$ (2.39)

and

$$\tau_{E,BSF,hi} = \frac{qn_i^2 W}{J_{0E}\Delta n} + \frac{qn_i^2 W}{J_{0BSF}\Delta n},$$ (2.40)

where J_{0E} and J_{0BSF} denote the emitter recombination current density and BSF recombination current density, respectively. The expressions given in 2.40 are only valid when the bulk lifetime is sufficiently long allowing generated carriers to reach both surfaces.

According to *Schmidt* [28], with a simplified surface recombination velocity, S_{eff}, defined as

$$S_{eff} \equiv \frac{U_s}{\Delta n},$$ (2.41)

the recombination rate can be found by combining Eq. 2.37 and Eq. 2.41 to

$$U_s = S_{eff}\Delta n = \frac{np}{qn_i^2}J_{0E}. \tag{2.42}$$

After *Lindholm and Sah* [53], *Neugroschel et al.* [54], and *Jain and Muralidhan* [55] (*cf.* also [56, 57]), a quasi-static approximation for the emitter has been developed. This approximation is taking into account the effect of the carriers in the emitter on the open circuit voltage decay:

$$\text{for } n\text{-type silicon} \qquad S_{eff} = \frac{N_D + \Delta n}{qn_i^2}J_{0E} \tag{2.43}$$

and

$$\text{for } p\text{-type silicon} \qquad S_{eff} = \frac{N_A + \Delta n}{qn_i^2}J_{0E}. \tag{2.44}$$

2.1.8. Influence of recombination channels on the lifetime

Design and operation of a high-efficient solar cell has two major key-factors, the minimization of the recombination rate and thus the maximization of the absorption of photons with $E > E_G$. That means the objective is to collect the minority carriers before they are lost to recombination, hence possible ways for a generated electron-hole pair would be:

- A photon generates an electron-hole pair near the surface: the hole might recombine at the top surface.

- A photon generates an electron-hole pair in the emitter: the minority carrier is collected when it crosses the junction.

- A photon generates an electron-hole pair in the base: the minority carrier is collected when it crosses the junction.

- A photon generates an electron-hole pair close to the rear surface: the electron recombines in the base.

- A photon generates an electron-hole pair near the rear surface: a minority carrier generated close to the rear surface might recombine at the rear surface.

The process of recombination, whereby electron-hole pairs are lost, sets energy free. This released excess energy is given to either a photon or phonon. Recombination therefore reduces the maximum achievable solar cell performance, in particular it effects the V_{oc} and J_{sc}. All paths of recombination discussed in this chapter may occur at the same time in a sample, but with merely different impact: for photovoltaic devices, some of those recombination processes do not occur or contribute negligibly to the total recombination rate, particularly radiative recombination. In Fig. 2.6, the effective lifetime dominated by the different recombination channels occurring in a photovoltaic device as a function of doping concentration for a p-type c-Si wafer is shown. One can see, that the impact of those recombination channels depend tremendously on the injection level or excess carrier density: for example, the Auger recombination dominates much more distinctive at high injection levels, whereas SRH recombination is the limiting-process below $1 \cdot 10^{16}$ cm^{-3}.

2.2. Characteristics of solar cells

Solving the minority-carrier diffusion equation with appropriate boundary conditions, the basic current-voltage (I-V) characteristic of a solar cell can be derived, which will be discussed in the following, *cf. Goetzberger et al.* [34]. A current density quoted in this work will be, unless stated differently, denoted as J, whereas I refers to a current flow.

2.2.1. Dark-current and open-circuit voltage

The characteristics of an ideal diode can be described with the diode equation, an expression for the current density at a *p-n* junction through a diode as a function of voltage. Thereby, J_{dark} denotes the diode leakage current density in the absence of light, and flows across a ideal solar cell under an applied voltage or bias, V. It occurs when a load is present between the two contacts

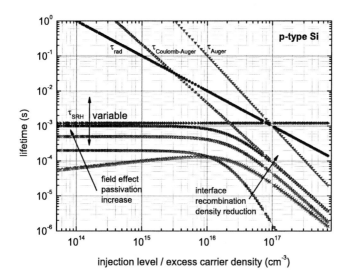

Figure 2.6.: Lifetime as a function of injection level (excess carrier density). Impact of SRH recombination (τ_{SRH} of 1 ms is assumed, which is arbitrary, depending on defect-type and -concentration), radiative recombination (calculated with $B = 1 \cdot 10^{-14}$ cm^{-3}/s), Auger recombination of free particles (calculated with $C_n = 0.99 \cdot 10^{-31}$ cm^6/s), and the Coulomb-enhanced Auger recombination (calculated with $\tau_{CA} = 2.374 \cdot 10^{24}$ $p_0^{-1.67}$) is shown. The effective lifetime is limited by SRH recombination due to deep defects. At high injection levels the effective lifetime is dominated by Auger recombination. After [45].

of a cell and hence the resulting potential difference generates a current which acts in reverse direction to the photocurrent. The dark current density, J_{dark} can be then defined for a ideal diode as

$$J_{dark} = J_0 \left(e^{\frac{qV}{k_B T}} - 1 \right),\tag{2.45}$$

where k_B is the Boltzmann constant and T represents the temperature in Kelvin [31]. The expression given in Eq. 2.45 is know as ideal diode law, where

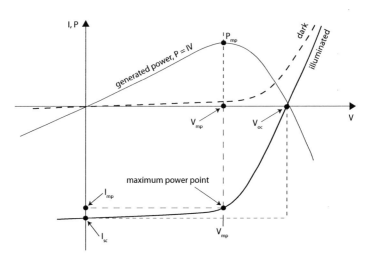

Figure 2.7.: Typical current-voltage characteristic and power output of a solar cell under dark conditions and under illumination.

J_0, as for as it concerns this thesis, denotes the saturation current density

$$J_0 = A \left(\frac{qD_n n_i^2}{L_n N_A} + \frac{qD_p n_i^2}{L_p N_D} \right),$$ (2.46)

with A denoting the cross-sectional area of the diode, D_n and D_p define the ambipolar diffusity, and L_n and L_p the electron and hole diffusion length, respectively. The diffusion length is given by $L_n = \sqrt{D_n \cdot \tau_n}$ and $L_p = \sqrt{D_p \cdot \tau_p}$ for electrons and holes, respectively. Fig. 2.7 shows the typical I-V characteristics and corresponding power output of a solar cell under dark conditions and under illumination. In case the terminals of the solar cell are isolated, the potential difference has its maximum value, hereafter denoted as open-circuit voltage

$$V_{oc} = \frac{nkT}{q} \ln \left(\frac{I_{sc}}{I_0} + 1 \right).$$ (2.47)

where J_{sc} denotes the short-circuit current density. The value for J_{sc} is obtained if the solar cell is short circuited so that there is no voltage at the cell.

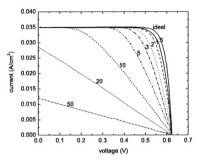
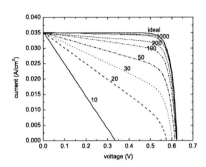

(a) Effect of increasing series resistance.   (b) Effect of reducing parallel resistance.

Figure 2.8.: Effect of series and parallel resistances. The outer curve in each case represents $R_s = 0$ and $R_{sh} = \infty$. The effect of the resistances reduce the area of the maximum power rectangle compared to $J_{sc} \cdot V_{oc}$.

2.2.2. Influence of series and parallel resistances

In case of heterojunction solar cells, the series resistances, R_s, arise from the resistance of the cell material to current flow, particularly through the front surface (a-Si:H(i), doped a-Si:H, TCO) to the contacts, and from resistive metal contacts. The main impact of a rising series resistance is to reduce the fill factor, although excessively high values may also reduce the J_{sc}. The effect of the series resistance on the I-V curve is illustrated in Fig. 2.8.

Power losses caused by the presence of a shunt or parallel resistance, R_{sh} or R_p, provide an alternate current path for the photo-generated current. Such a diversion reduces the amount of current flowing through the cell and reduces the V_{oc}. The major part of an existing shunt resistances is determined by leakage currents along the edges of a solar cell. Point-defects, such as interruptions of the p-n junction by applying an excessively thin a-Si layer, may also lead to low parallel resistance [2, 34]. Typical values for area-normalized shunt resistances are in the MΩcm^2 range [31]. The influence of shunt resistances on solar cell parameters is shown in Fig. 2.8. As seen in Fig. 2.8, V_{oc} and J_{sc} are only affected for excessive values of R_s and R_{sh}, but high series and/or low parallel resistances reduce the fill factor; in turn, for an efficient solar cell R_s should be as small as possible and R_{sh} as large as possible.

2.2.3. Fill factor

As presented in Fig. 2.7, the maximum power point, P_{mp}, determined by the cell's operating point, defines the maximum voltage V_{mp} and hence the maximum current I_{mp}. The *fill factor* (FF) describes the 'squareness' of the IV-curve, shown in Fig. 2.7, and can be defined as the ratio

$$FF = \frac{J_{mp} V_{mp}}{J_{sc} V_{oc}}.\tag{2.48}$$

The FF can be determined also as a function of series resistance. As can be found in Ref. [2, 32], the normalized open circuit voltage is $v_{oc} = V_{oc}/(n_{id1}(k_B b/q))$, and the normalized resistances are $\rho_s = R_s/R_{ch}$ and $\rho_p = R_{sh}/R_{ch}$, with the characteristic resistance, $R_{ch} = V_{oc}/J_{sc}$. The fill factor is limited by the diode ideality factor n_{id1} as

$$FF_0 = \frac{v_{oc} - ln\,|v_{oc} + 0.72|}{v_{oc} + 1}.\tag{2.49}$$

Assuming that the open-circuit voltage and short-circuit current are not affected by the series resistance, the impact of R_s on FF can be determined as

$$FF_{RS} = FF_0 \left(1 - \rho_s\right).\tag{2.50}$$

An empirical equation for the FF including the impact of the parallel resistance is given by

$$FF = FF_0 \left(1 - \rho_s\right) \left(1 - \frac{v_{oc} + 0.7}{v_{oc}} FF_0 \frac{1 - \rho_s}{\rho_p}\right).\tag{2.51}$$

2.2.4. Efficiency

The *efficiency* of a solar cell can be defined as the ratio of power density delivered at the P_{mp} to the incident light power density, P_s, which is usually corresponding to standard test conditions (25 °C, 1000 W/m^2, spectrum AM1.5g [1]).

---

[1] AM1.5g denotes the standard solar spectrum in case the light passed an air mass of 1.5 times the distance of the sun in zenith. 'g' corresponds to the 'global' spectrum considering direct and diffuse radiation.

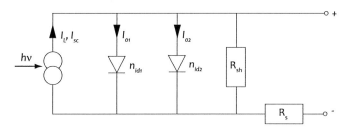

Figure 2.9.: Equivalent circuit for the two-diode model of a real solar cell including series and shunt resistances. The solar cell is depicted as a diode with ideality $n_{id1} = 1$ and a current saturation I_{01}. The second diode with an ideality factor n_{id2} models recombination within the space charge region, the series resistance and shunt resistance account for electrical losses.

$$\eta = \frac{J_{mp}V_{mp}}{P_S},$$ (2.52)

or related to J_{sc} and V_{oc} using FF, by

$$\eta = \frac{J_{sc}V_{oc}FF}{P_S}.$$ (2.53)

2.2.5. Equivalent circuit of a real solar cell

The most commonly accepted model describing solar cells is the two diodes model; it schematically illustrates the equivalent circuit diagram for a real solar cell, as shown in Fig. 2.9, after [30]. The solar cell is characterized by a diode with an ideality factor[2], n_{id1}, and a current saturation, I_{01}. The current saturation, I_{01}, for a p- or n-type basis is found to be

$$I_{01} = \frac{qD_{n,p}n_i^2}{N_{A,D}L_{eff}},$$ (2.54)

with a doping level of N_A or N_D, and a diffusion constant D_n or D_p, respectively, and an effective diffusion length, L_{eff}. The recombination in the space charge region W is reflected by a second diode characteristic resulting in a diode ideality factor, n_{id2}, and a recombination current, I_{02}.

---

[2]The ideality factor describes the deviation of how closely the diode follows the ideal diode equation. The ideality factor n typically lies between 1 and 2.

The current-voltage characteristic of the two-diode-model incorporating the parasitic resistances can be expressed as

$$I(V) = I_{01} \left[exp \left(\frac{e\,(V - IR_s)}{n_{id1}kT} \right) - 1 \right] + I_{02} \left[exp \left(\frac{e\,(V - IR_s)}{n_{id2}kT} \right) - 1 \right]$$
$$+ \frac{V - IR_s}{R_{sh}} - I_{sc}.$$

(2.55)

I_{02} specifies the dark saturation current of the depletion region, which can be defined as

$$I_{02} = \frac{eW\sigma_e N_T v_{th} n_i}{2},$$

(2.56)

where W is the width of the depletion region, σ_e is the capture cross section for electrons, N_T is the trap density, and $v_{th} = \sqrt{3kT/m_n^*}$ denotes the thermal velocity. It has to be noted, that I_L is the light generated current inside the solar cell and hence the correct term to use in the solar cell equation. At short circuit conditions the externally measured current is I_{sc}. Since I_{sc} is usually equal to the absolute light current amount I_L, the two are used interchangeably. The currents in equation 2.55 are usually given in units of mA/cm^2, whereas the resistances are given in units of Ωcm^2. The magnitude of this current, disregarding all losses in the cell (recombination, optical), with AM1.5 radiation can reach a peak value of 44 mA/cm^2 [34]. In case of very high series resistance ($\geq$ 10 Ωcm^2) I_{sc} is less than I_L and writing the solar cell equation with I_{sc} would be incorrect [2]. In case the solar cell is operated under open circuit conditions, no current is flowing ($I = 0$) and from equation 2.55 the open circuit voltage can be defined as

$$V_{oc} = \frac{2kT}{q} \cdot ln \left\{ \left[\left(\frac{I_{02}}{2I_{01}} \right)^2 + \frac{I_L + I_{01} + I_{02}}{I_{01}} \right]^{0.5} \cdot \frac{I_{02}}{2I_{01}} \right\},$$

(2.57)

or, for the case of open circuit conditions, neglecting the influence of the second diode and the parallel resistance, V_{oc} can be written as

$$V_{oc} = n_{id1} \frac{k_B T}{q} ln \left| \frac{I_{sc}}{I_{01}} + 1 \right|.$$

(2.58)

Figure 2.10 illustrates the impact of the two-diode-parameters I_{01}, I_{02}, R_s, and R_p on V_{oc} and FF, respectively, calculated by the two-diode-model for various parameter combinations, *cf*. [58]. Typical values for those two-diode-parameters for high efficiency solar cells are $I_{01} \approx 1 \cdot 10^{-13}$ A/cm^2, $I_{02} \approx 1 \cdot 10^{-9}$ A/cm^2, $R_s \approx 0.5$ Ωcm^2, and $R_{sh} \geq 1 \cdot 10^6$ Ωcm^2. The impacts of those parameters are described in more detail in *Glunz* [59].

2.3. Amorphous / crystalline silicon (a-Si:H/c-Si) heterojunction solar cells

In prospect of a-Si:H/c-Si heterojunction solar cells investigated in this work, a brief discussion about these types of cells will be given, followed by optical and electrical properties which will be described within the context of the electronic band theory of solids, and the nature of the amorphous material used for fabrication.

2.3.1. Properties of amorphous silicon

Crystalline silicon (c-Si) can be designated as four-fold coordinated network, whereas one silicon atom is tetrahedrally bonded to four neighboring silicon atoms. This tetrahedral structure is continued, forming a well-ordered lattice (crystal). In contrary, within amorphous silicon (a-Si), not all the atoms are four-fold coordinated due to a continuous random network. Due to the disordered nature of the amorphous silicon some atoms feature an unsaturated band (*'dangling bond'*) (db). These dangling bonds can be described as broken covalent bonds, which are also found on the surface of crystalline silicon due to the absence of lattice atoms above them. The amorphous character yield either a neutral, positive or negative charge of the db. The deviation of the bond-angles averages $\pm 10°$ [60]. As a result, once i.e. a metal is deposited on a silicon surface, these dangling bonds rise to interface states within the energy bandgap of silicon. The bonding structure of both crystalline and amorphous silicon is depicted in Fig. 2.11.

If desired, the material can be passivated by hydrogen. Additional hydrogen bonds to the db and can reduce the db density by several orders of mag-

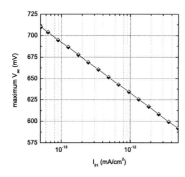

(a) Impact of I_{01} on V_{OC} using $I_{02} = 1 \cdot 10^{-18}$ A/cm^2, $R_s = 0$ Ωcm^2, $R_{sh} = 1 \cdot 10^{10}$ Ωcm^2, $I_L = 42$ mA/cm^2, $n_{id1} = 1$, $n_{id2} = 2$.

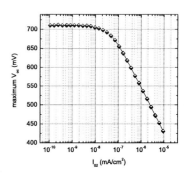

(b) Impact of I_{02} on V_{OC} and FF using $I_{01} = 5 \cdot 10^{-14}$ A/cm^2, $R_s = 0$ Ωcm^2, $R_{sh} = 1 \cdot 10^{10}$ Ωcm^2, $I_L = 42$ mA/cm^2, $n_{id1} = 1$, $n_{id2} = 2$.

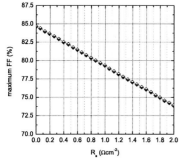

(c) Impact of R_s on FF using $I_{01} = 5 \cdot 10^{-14}$ A/cm^2, $I_{02} = 1 \cdot 10^{-18}$ A/cm^2, $R_{sh} = 1 \cdot 10^{10}$ Ωcm^2, $I_L = 42$ mA/cm^2, $n_{id1} = 1$, $n_{id2} = 2$.

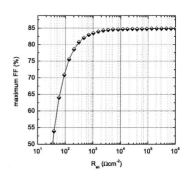

(d) Impact of R_{sh} on FF using $I_{01} = 5 \cdot 10^{-14}$ A/cm^2, $I_{02} = 1 \cdot 10^{-18}$ A/cm^2, $R_s = 0$ Ωcm^2, $I_L = 42$ mA/cm^2, $n_{id1} = 1$, $n_{id2} = 2$.

Figure 2.10.: Limitation of the solar cell performance calculated using the two-diode-model. After *Kray* [58].

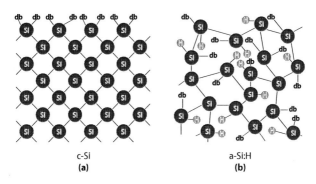

c-Si
(a)

a-Si:H
(b)

Figure 2.11.: (a) 2-dimensional schematic view of crystalline (c-Si) and (b) hydrogenated amorphous (a-Si:H) silicon. Silicon dangling bonds (db) constitute a broken covalent bond in a-Si:H, and are also found on the surface of c-Si due to the absence of lattice atoms above them. The ratio of db to saturated bonds in this figure is not to scale.

nitude. Thereby, SiH_n groups are generated (mostly SiH and SiH_2), forming a-Si:H. High-quality hydrogenated amorphous silicon (a-Si:H) exhibits a hydrogen content of around 7-13 % (*cf. Street* [60], *Janssen et al.* [61]), leading to a sufficiently low amount of defects within devices. Hereby, the density of states in the bandgap is reduced from $1 \cdot 10^{19}$ cm^{-3} for a-Si (*cf.* [60]) down to $1 \cdot 10^{16}$ cm^{-3} for a-Si:H(i). Thus, the 'band-tails' become precipitous and the bandgap is expanded, after *Tauc* [62], from 1.55 eV for a-Si to around 1.7 eV for a-Si:H(i). In Fig. 2.12 the corresponding model of state densities after *Davis and Mott* [63] is shown, denoted as Mott-CFO model (after *Mott, Cohen, Fritzsche, and Ovshinsky, cf. Tauc* [62, pg. 110]). It has to be mentioned that this model appears to be a simplified model and not consistent with the present day theory. Material properties and corresponding electronic properties can be found elsewhere *Joannopoulos et al.* [23], *Street* [60], *Kanicki* [64]. The main predominance of a-Si compared to c-Si relies in its production technique: a very thin a-Si:H film can be deposited onto large areas by i.e. plasma enhanced chemical vapor deposition (PECVD), while doping is achieved during plasma deposition by decomposition of doping gases to form either p- or n-type layers. The resulting amorphous or micro-crystalline (μc-Si:H) network, and hence the optical and electrical properties of the so deposited layers, are

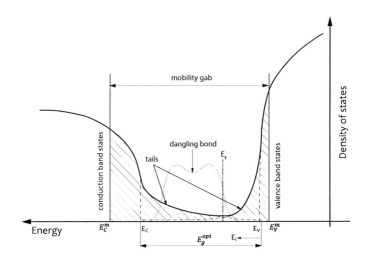

Figure 2.12.: Density of states as a function of energy E in a-Si:H. The exponential
slope of the density of states are denoted 'band-tails'. The valence band-
tail, with a characteristic decay-constant of 50 meV (unlike the conduction
band-tail with 30 meV)), determines the sub-bandgap absorption with the
Urbach energy, *cf*. [62]. E_g^{opt} is determined by extrapolation of the delo-
calized states, *cf*. section 3.2.3. E_V^m and E_C^m are the mobility edges. The
mobility gap contains localized states in between the bandgap, while loco-
motive states outside the bandgap determine mainly the carrier transport.
Open bonds lead to deep defect states in the middle of the bandgap.

adjusted by process parameters as explained in section 3.1.1. A hydrogen
precursor for the a-Si:H can be silane (SiH_4) or additional hydrogen (H_2).

2.3.2. a-Si:H/c-Si heterostructures - band diagram and band discontinuities

Once an *n*-type and an *p*-type semiconductor material are joined, a heteroint-
erface, or *p-n* junction is formed. A *p-n* junction aggregates carrier generation,
recombination, diffusion and drift effects into a single device. It happens to
be that isotypes (*n-n* or *p-p*) and anisotypes (*n-p* or *p-n*) of heterojunctions

exist. The a-Si:H/c-Si heterostructures in this work are anisotypes, therefore the following section covers this type of heterojunctions.

The simplest model describing those heterointerfaces for isolated semiconductors of opposite types and at thermal equilibrium was established by *Anderson* [65]; a corresponding energy-band diagram is shown in Fig. 2.13. In this case, the two semiconductors represent a-Si:H (wider E_G) and c-Si (smaller E_G) and are assumed to have different work functions Φ_m, and different electron affinities χ. Once a junction is formed between these two semiconductors, an exchange of charge carriers is proceeding till a thermal equilibrium is reached. The Fermi level must coincide on both sides in equilibrium.

Since the p-type region has a high hole concentration and the n-type region a high electron concentration, electrons and holes diffuse to the opposite side. However, when the electrons and holes in the p-n junction move to the other side of the junction, they leave behind exposed charges on dopant atom sites, which are fixed in the crystal lattice and are unable to move. Positive and negative ion cores are exposed on the n-type side and the p-type side, respectively. An electric field $\vec{E}$ is formed between the positive ion cores in the n-type material and negative ion cores in the p-type material; forming the 'depletion region' (since the electric field quickly sweeps free carriers out, hence the region is depleted of free carriers) [2]. The total 'built-in' potential Ψ_{bi} is equal to the sum of the partial built-in voltages ($\Psi_{b1}+\Psi_{b2}$) and is defined by the difference of the work functions by $\Psi_{bi} = 1/q(\Phi_{m1} - \Phi_{m2})$. The work function and electron affinity are defined as the energy required to remove an electron from the Fermi level E_F and from the bottom of the conduction band E_c, respectively, to a position just outside the material (vacuum level) [66]. The conduction band edges of the two semiconductors merge at the interface, which forms the conduction band discontinuity

$$\Delta E_c = \chi_2 - \chi_1. \tag{2.59}$$

The sum of conduction and valence band discontinuities correspond to the difference of the bandgaps, so that the valence band discontinuity can be ob-

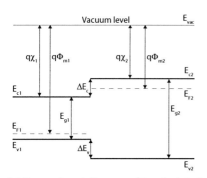

(a) Energy-band diagram of two isolated semiconductors; of opposite types and different E_G. The smaller E_G is p-type.

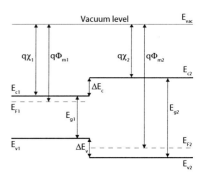

(b) Energy-band diagram of two isolated semiconductors of opposite types and different E_G. The smaller E_G is n-type.

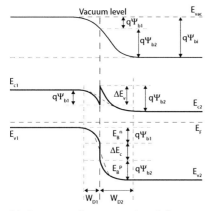

(c) Corresponding energy-band diagram of their idealized anisotype heterojunction at thermal equilibrium.

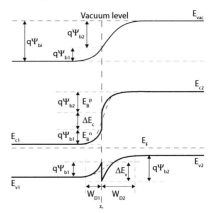

(d) Corresponding energy-band diagram of their idealized anisotype heterojunction at thermal equilibrium.

Figure 2.13.: Band diagram after *Anderson* [65]. χ defines the electron-affinities, Φ the work function, and E_g the bandgap of the semiconductor. Due to the difference in the bandgap of the semiconductors, ΔE_v and ΔE_c illustrate the valence band discontinuity and the conduction band discontinuity, respectively. E_B describes the corresponding band bendings, and the activation energy E_A is given by $E_A = q\Psi_{bi}$. Ψ_{bi} is the build-in voltage, which can be calculated by $\Psi_{bi} = 1/q(\Phi_{m1} - \Phi_{m2})$. W_{D1} and W_{D2} define the comprehensivenesses of the space charge region in the semiconductor materials. The diagrams are not to scale in respect of shown bandgaps.

tained by

$$\Delta E_v = \left(E_{g2} - E_{g1}\right) - \Delta E_c = \left(E_{g2} - E_{g1}\right) - (\chi_2 - \chi_1). \tag{2.60}$$

Eq. 2.60 can be merged to

$$\Delta E_c + \Delta E_v = E_{g2} - E_{g1}. \tag{2.61}$$

The discontinuities directly depend on the charge carrier density; resulting from doping concentration and defect levels. The higher the charge carrier density of one material compared to the other material appears, the smaller are the band discontinuities.

An overview of heterostructures is given in detail in *Milnes and Feucht* [67] and *Sharma and Purohit* [68]. Whereas the thermodynamic model established by *Anderson* [65] turned out to be not quite precise (deviation of ± 0.15 eV regarding the occurring band discontinuities from theory to experimental results), various attempts have been made to describe the heterojunction in more detail:

- *Frensley and Kroemer* [69] determined the band structure of semiconductors in the volume and extrapolated the valence band discontinuity from the position of the valence band maxima.

- *Adam and Nussbaum* [70] established an enhanced model of the *Anderson* model using the intrinsic level of the semiconductor as the reference potential.

- *von Ross* [71] proposed a model using the conduction band as reference potential. The valence band discontinuity is extrapolated from the bandgap differences.

- *Harrison* [72] described a model using the LCAO theory (Linear Combination of Atomic Orbitals). The valence band discontinuities are extrapolated from the difference of the energetic levels of the valence band maxima.

- *Tersoff* [73] presented a model determining the band discontinuities from interface properties; the model assumes that the interface of a semiconductor can be assigned to a neutral level.

2.3.3. Optical losses in solar cells

For most silicon solar cells, light of the entire visible spectrum has enough energy to create electron-hole pairs and therefore all visible light would ideally be absorbed. Optical losses consist of light which could have generated an electron-hole pair, but does not for the reason that once an incident photon hits the surface of a silicon wafer, it will be either absorbed in the material, or reflected from the top surface, or transmitted through the material. For photovoltaic devices, optical losses occur due to the latter two processes, as photons which are not absorbed do not generate electron-hole pairs. Optical losses mainly have an impact on the power output of a solar cell by lowering the short-circuit current. Figure 2.14 shows a schematic drawing of a silicon based heterojunction solar cell, indicating the main optical loss mechanisms: (i) reflection at the top surface of the solar cell, (ii) shading losses of several percent due to reflection at the front metal contacts, and (iii) transmission of photons through the solar cell in the long wavelength region, *cf*. Fig. 2.15.

There are a number of mechanisms to reduce the optical losses, such as:

- Minimizing the top contact coverage of the cell surface.

- Reduction of the reflection by anti-reflection coatings on the top surface of the cell, such as transparent conductive oxides (TCO).

- Reduction of the reflection by surface texturing, as bare silicon is quite reflective (about 36 % reflactance weighted with the AM1.5g spectrum, [77]). Surface texturing by i.e. randomized pyramids on the silicon wafer surface enables a second or third chance of an incident photon to enter the substrate and generate an electron-hole pair, depending on the benignity of the texturization quality.

- Thickness increase of the solar cell base material to increase absorption, although any light which is absorbed more than a diffusion length away

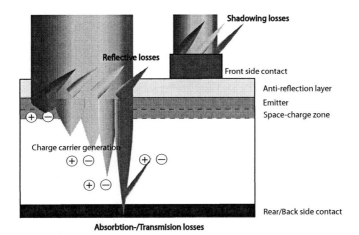

Figure 2.14.: Sketch of possible optical losses in a heterojunction solar cell, mainly
affecting the J_{sc}. The losses occur mainly due to reflection at the top
surface, shadowing effects of the metal contacts, and transmission in
the long wavelength region. Since silicon is an indirect semiconductor,
the absorption probability for photons with low energy ($\lambda > 1000$ nm)
decreases. After [74].

from the junction will not typically contribute to short-circuit current
since the carriers recombine.

- Increase of the optical path length in the solar cell by a combination of
  surface texturing and light trapping [2].

Recently, *Swanson* calculated the limit efficiency for silicon solar cells, stating that a silicon solar cell can not overcome an efficiency of 29 %. *Swanson*
pointed out that from the ideal efficiency of 29 %, a textured front surfaces
with a single anti-reflection-coating (ARC) will have a weighted reflectance of
around 2 %, which lowers the ideal efficiency to 28.2 %. In addition, the back
surface reflactance can not be 100 %, but it can easily be 90 %, lowering the ideal
efficieny to 27.8 %. The correspoonding J_{sc} for perfect optics would be 42.5
mA/cm^2, including a 2 % front reflactance it decrease to 41.6 mA/cm^2, and including a decreased back reflactance it would decrease to 90 % (41.1 mA/cm^2).

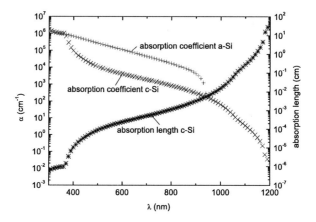

Figure 2.15.: Absorption coefficient (α) of c-Si [75] and a-Si:H(i) [76] and corresponding absorption length of c-Si as a function of wavelength. The absorption length defines the distance where the intensity is reduced to 37 % of its original power.

Summarizing, going from ideal optics to readily achievable optics, the relative efficiency decrease can be estimated as high as 3 % efficiency.

2.3.4. a-Si:H/c-Si heterojunction solar cells

Heterojunction solar cells consisting of doped amorphous silicon directly deposited onto the c-Si substrate results usually in a low open-circuit voltage (V_{oc}) and fill factor (FF) compared to conventional *p-n* diffused solar cells. These poor properties appeal to be caused by the recombination process throughout the depletion region at the a-Si/c-Si heterojunction. Moreover, the a-Si has mid-gap states that possibly increase the leakage current by a tunneling process [6]. To suppress these tunneling processes a spacer containing less density of trap levels can be inserted. Based on the ideas of *Taguchi* [79], *Wakisaka et al.* [80], and *Tanaka et al.* [8], the main feature of the HIT™(Heterojunction with Intrinsic Thin Layer) concept developed by *SANYO* consists of a very thin intrinsic amorphous silicon layer between p$^+$-type a-Si

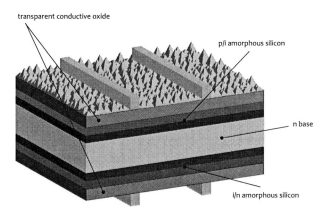

transparent conductive oxide

p/i amorphous silicon

n base

i/n amorphous silicon

Figure 2.16.: Schematic diagram of a standard heterojunction solar cell. The cell is composed of randomly structured c-Si substrate sandwiched between p^+/i a-Si layers on the illuminated side and i/n^+ a-Si layers as BSF. Transparent conductive oxide layers and metal electrodes are formed on both doped layers. The textured surface of the crystalline wafer and the amorphous layer is not shown for simplicity. After *Taguchi et al.* [9].

and n-type c-Si. Figure 2.16 shows a cross-sectional view of a commercially available HIT solar cell.

The *p-n* junction is realized by PECV deposition of non-doped (intrinsic) and p-type doped a-Si layers on an n-type c-Si substrate. The back-surface-filed on the backside of the device is structured with stacked non-doped and n-type a-Si layers. The transparent conductive oxide (TCO) is sputtered on both doped a-Si layers and contacted with metal electrodes (after [6]). The fundamental requests of the heterojunction solar cell concept are:

- High efficiency, cost-effective approach with a simple structure avoiding complicated structural techniques, such as partly heavy doping or a partial oxidation method. The high efficiency might be credited to the effective carrier trapping within the generation region (c-Si) utilizing the double heterostructure.

- Low plasma process features high-quality, low-defect amorphous silicon films.

- High V_{oc} compared to standard c-Si solar cells due to the low saturation current densities of the a-Si:H/c-Si hetero-contacts, *cf. Swanson* [78].

- Low thermal damage due to PECV deposition temperature below 200 °C, suppressing the initiation of intrinsic impurities located in the solar grade substrate, which would lead to a decrease of the effective carrier diffusion length. The degradation of the minority carrier lifetime for the substrate is negligible even for low-quality Si materials, *cf. Taguchi et al.* [6].

- Excellent surface passivation of the c-Si surface defect states at the front and backside by application of intrinsic a-Si layers. This enhances the open-circuit voltage, *cf. Taguchi et al.* [6], *Page et al.* [81].

- Improved high-temperature performance compared to conventional c-Si solar cells, *cf. Maruyama et al.* [82]. Conventional c-Si based solar cells have a relatively poor high-temperature performance compared with a-Si cells. Heterojunction solar cells, consisting of both a-Si and c-Si, exhibit an improved temperature performance, *cf. Taguchi et al.* [6].

- Increased stability: The Staebler-Wronsky effect, which is seen on a-Si based solar cells, does not occur to heterojunction solar cells, *cf. Taguchi et al.* [6]. This might be due to the fact that the layer is very thin.

2.3.4.1. Requirements for high-efficiency heterojunction cells

Within the concept of high-efficient heterojunction solar cells, special requirements are devoted to:

Surface passivation The influence of the intrinsic intermediate a-Si:H layer on the V_{oc} was first investigated at the Material Research Center of *SANYO* by *Sawada et al.* [7], *Tanaka et al.* [8], *Tsuda et al.* [83]. An increased V_{oc} compared to cells without a-Si:H(i) and reduced saturation current density by two orders of magnitude could be attributed to the application of a-Si:H(i). The improvements can be explained due to a decreased defect state density at the

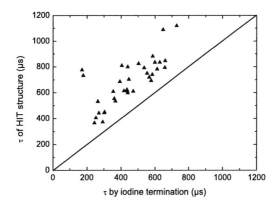

Figure 2.17.: Carrier lifetime realized by HIT structures, compared to iodine termination. After [82].

a-Si:H(i)/c-Si interface. In Fig. 2.17 the carrier lifetime tendency estimated by μ-WPCD (described in section 3.6.1) of various wafers with chemical passivation by iodine termination[3] and with surface passivation by a-Si:H(i) first reported by *Taguchi et al.* and *Maruyama et al.* are shown. They concluded that the a-Si:H(i) passivation results in a better surface passivation of c-Si compared to conventional thermal passivation.

Wolf [85] posted that surfaces could have zero recombination velocity by using heterojunction contacts. A perfect minority carrier mirror for silicon is not practically accomplished at present. However, it is being approached by the implementation of amorphous silicon contacts used in *SANYO* HIT cells ([78]). In addition, *Taguchi et al.* [9] stated that by optimizing the deposition conditions of the a-Si:H(i), the surface recombination velocity is estimated to be less than 100 cm/s for their cells. Therefore, the HIT cell has a sufficiently low interface state density. *Taguchi et al.* also pointed out that there would be no need to consider the optimum band offset affected by the interface state density, but the way to reduce the band offset to block the minority carriers.

---

[3]Iodine passivation is usually considered featuring better surface passivation qualities than thermal oxidation, *cf.* [84].

High-quality a-Si:H films   *Taguchi et al.* [6] also reported the correlation between the defect density (N_d) of the deposited a-Si:H(i) films at various substrate temperatures as a function of the deposition rate and optical bandgap,
cf. Fig. 2.18. Both the deposition rate and the optical bandgap were adjusted
by the processing parameters such as gas flow rate, deposition pressure and
deposition power density. *Maruyama et al.* concluded that N_d is mainly depending on the optical gap rather than on individual deposition conditions.
For their deposition conditions, an optimum region of optical gap for lower N_d
exists. A high-quality a-Si:H film considered to be suitable for high-efficiency
solar cells would be the one with the lowest N_d.

Optimization   Stated by *Maruyama et al.*, achieving higher conversion efficiency demands the improvement of V_{oc}, J_{sc} and FF at the same time. A higher
V_{oc} might be achieved by a further decrease of the surface recombination velocity at the a-Si:H/c-Si interface. Key approaches contain (i) an enhanced
cleaning process of the c-Si surface prior deposition of the a-Si:H films, (ii)
the optimization of the band offset, (iii) a decrease of surface recombination
by high quality a-Si:H, and (iv) lower plasma and/or thermal damage to c-Si
surface to increase electric conductivity. The impact of an enhanced cleaning
process and a low damage deposition of a-Si:H on the V_{oc} of *SANYO* HIT cells
was shown by *Maruyama et al.*, depicted in Fig. 2.19. It is essential to optimize not only film-qualities of the a-Si:H films, but also plasma and thermal
damage to the c-Si surface during a-Si:H deposition. The finding that solar cell
optimization is ineffective if optimization schemes focus on a single aspect of
the device at a time is also confirmed by *von Roedern* [20].

2.3.4.2. *n-p* or *p-n* heterojunction structure?

The only commercial available and at the same time technological highest
developed a-Si:H/c-Si heterojunction solar cells are realized by *SANYO*. Many
groups are working trying to gain better understanding of the principles of
those cells (i.e. *Stangl et al.* [13], *Rostan et al.* [14], *Wang et al.* [16], *Tucci et al.*
[17], *Fujiwara and Kondo* [18], *Olibet et al.* [86]); almost all of them working

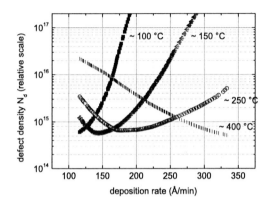

(a) Defect density (N_d) of a-Si:H films as a function of the deposition rate.

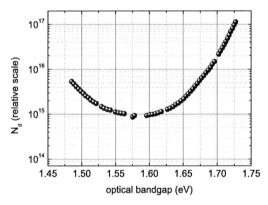

(b) N_d of a-Si:H films as a function of the optical bandgap.

Figure 2.18.: Defect density (N_d) of a-Si:H films. After [6].

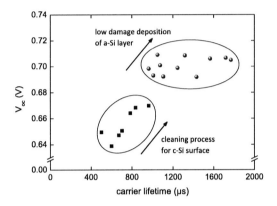

Figure 2.19.: Correlation of V_{oc} and carrier lifetime of the HIT™solar cell after a-Si:H deposition. After [82].

with n-type and p-type wafer substrates; in contrary *SANYO* realizes its cells explicitly on n-type wafer material.

Figure 2.20 illustrates the transport mechanisms for the charge carriers at an illuminated heterostructure for both a-Si:H(p)/c-Si(n) and a-Si:H(n)/c-Si(p). Those mechanism help to understand whether n-type or p-type material is preferable. It is obvious that the charge carriers have to overcome the band discontinuities. For a junction with n-doped base, the electrons yielding from the emitter towards the base, and the holes towards the emitter, which are hampered due to the valence band discontinuity. That means for an a-Si:H (p)/c-Si (n) junction, the valence band discontinuity should be small. For an a-Si:H(n)/c-Si(p) junction, the electrons are hampered in their way to the emitter due to the conduction band discontinuity.

Froitzheim [87] simulated both opposite structures and concluded that for a a-Si:H(p)/ c-Si(n) structure the V_{oc} and J_{sc} are higher and the defects at the interface are decreased, whereas the only advantage of the a-Si:H(n)/c-Si(p) structure is an enhancement of the FF. Based on those knowledge, one can conclude that the a-Si:H(p)/c-Si(n) structure would be predominant if: (i) the recombination at the interface a-Si:H(p)/ c-Si(n) is suppressed more effective due to the

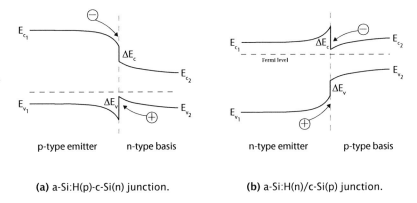

(a) a-Si:H(p)-c-Si(n) junction.   **(b)** a-Si:H(n)/c-Si(p) junction.

Figure 2.20.: Transport of charge carriers over the band discontinuities for a heterostructure in thermal equilibrium.

asymmetric level of the band discontinuities, (ii) the structure a-Si:H(p)/c-Si(n) exhibits a higher V_{oc} level due to the lower backward saturation current, (iii) the quantum efficiency for $\lambda > 600$ nm appears to be independently of the absorber doping type for higher diffusion-lengths in the c-Si absorber material. However, a-Si:H(p) used as an emitter features a high blue light response due to a higher minority carrier mobility. This leads to an increased J_{sc} value.

CHAPTER 3.

SAMPLE PREPARATION AND CHARACTERIZATION METHODS

In this chapter basic aspects on sample preparation methods and analysis techniques are discussed. The sample preparation section contains a brief description of general process steps for the fabrication of heterojunction solar cells and a discussion of the plasma enhanced chemical vapor deposition setup used in this work for deposition of a-Si:H and its alloys, such as a-SiO$_x$:H, a-SiC:H, and μc-Si:H. In addition, the characterization methods applied for investigation of the optical, electrical and morphological properties of plasma enhanced deposited a-Si:H/μc-Si:H layers are discussed regarding either 'as deposited' (onto glass or c-Si) or in terms of a complete solar cell structure. In particular, (i) the determination of the bandgap of a semiconductor by use of spectroscopic ellipsometry, (ii) the corresponding local vibrational modes analysis of amorphous and micro-crystalline layers by use of μ-Raman spectroscopy, (iii) cell characterization methods of heterojunction solar cells, as well as (iv) lifetime measurements using either quasi-steady-state-, transient- or microwave-photoconductance decay are described.

3.1. Preparation of a-Si:H/c-Si heterostructures

The fabrication of heterojunction devices consisting of a-Si:H/c-Si heterostructures is discussed in this section. In particular, the fabrication procedure will be described with emphasis on plasma deposition, whereas minor fabrication steps, such as (i) wafer pre-treatment, (ii) deposition of an transparent conductive oxide, and (iii) formation of metal contacts are described elsewhere (*cf*. Appendix A). The fabrication of heterojunction solar cells can be divided into the following process steps:

- *Cleaning process* - In order to create an excellent hetero-interface, it is essential to use a clean c-Si wafer surface; this can be achieved by the removal of metallic impurities, organic contaminations and surface films, such as native oxides and absorbed molecules (described in section A.1).

- *HF (hydrofluoric acid) dip* - prior deposition with the PECVD setup, it is essential to remove any native oxide on the silicon surface to obtain an excellent hetero-interface (refer to section A.1).

- *PECVD (plasma enhanced chemical vapor deposition)* - High-quality a-Si:H films should be deposited by low damage plasma process, *cf*. [9].

- *Annealing* - Post-annealing of deposited a-Si:H layers and its alloys is preferable in order to enhance dark conductivity.

- *TCO (transparent conductive oxide)*, serving as an anti-reflection coating and increasing the lateral conductivity; refer to section A.2.

- *Formation of metal contacts* by evaporation on front and back side, including photo-lithography, lift-off and masking processes, refer to section A.3.

Those process steps are summarized in the flowchart in Fig. 3.1.

3.1.1. High-frequency plasma enhanced chemical vapor deposition

Depositions of amorphous or micro-crystalline silicon layers, whether they are intended to be used as passivation-, emitter- or BSF-layer, or for material characterization, are performed in a commercial, parallel plate, capacitively coupled three chamber Plasma Enhanced Chemical Vapor Deposition (PECVD) system (*Material Research Group*). In this technique a plasma occurring during the decomposition of the gaseous precursors, which can be i.e. SiH_4, H_2, CH_4, CO_2, and dopant gases such as PH_3 and TMB, is used. During the decomposition inelastic collisions between high-energetic electrons and the gaseous precursor atoms result in a dissociation into atomic and ionic species. Figure 3.2 shows a schematic of one plasma chamber used in the present work.

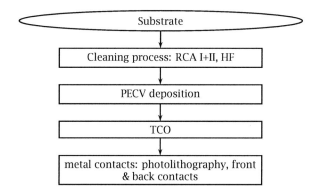

Figure 3.1.: Flowchart illustrating sample preparation of a-Si:H/c-Si heterostructures.

The pathways for the chemical reactions of SiH_4 and its plasma products occurring during the operation of PECVD systems can be found in i.e. [88, 89].

In general, the samples are placed into a 10×10 cm squared sample holder, which is suitable for up to 4 inch wafer substrates, and transfered via a load lock into one of the three chambers (each chamber for either intrinsic, p-doped or n-doped layers to prevent contamination). The sample holder is attached to the upper (electrically grounded) electrode, while the rf-power is capacitively coupled to the lower electrode. A plasma power as high as 100 W can be adjusted. For the calculation of the plasma power in W/cm^2, the area of the lower electrode with a radius of $r_e = 6.75$ cm is used. The rf-PECVD system operates either at a fixed frequency of 13.56 MHz, or is driven by a second rf-generator, at frequencies ranging from 13.56 MHz up to very high frequencies (VHF) at 110 MHz. Changing the excitation frequency does not necessarily lead to higher deposition rates, but possible changes in the micromorph structure are likely.

The distance between the parallel electrodes (d_e) affects the emerging network of the amorphous or micro-crystalline deposited layers, and have to be adjusted precisely. Whereas the distance between the electrodes (d_e) in the chamber reserved for intrinsic layers only are set to 19 mm, so that a rather amorphous network can be achieved. The d_e in both n-type and for p-type

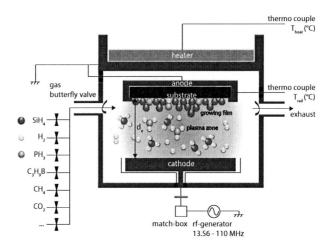

Figure 3.2.: Schematic view of a parallel plate, capacitively coupled PECVD chamber used for depositions in this work. The samples are loaded via a load lock into one of the three chambers (each chamber for either intrinsic, p-doped or n-doped layers to prevent contamination). The sample holder is attached to the upper electrically grounded electrode, while the rf-power is capacitively coupled to the lower electrode. The distance between the electrodes is denoted as d_e. A thermo couple sets the heater temperature (T_{heat}) and the rail temperature (T_{rail}). The latter matches in ideal case the sample temperature (T_{sample}) during deposition. The gaseous precursors are subject to change depending on the experimental needs; a complete list of available and used gases is given in the appendix (section A.4). Those gaseous precursors are introduced to the chamber via mass-flow controllers (MFC). A decomposition of SiH_4, PH_3, and H_2 is depicted exemplarily: SiH_3 and SiH_2 species are dissociated in the rf-field. The dimensions are not to scale.

doping chambers is set to 12 mm, in order to obtain rather micro-crystalline films than amorphous. Prior to deposition, the samples are heated up before purging with Argon gas. The Ar pressure is typically 5 mTorr for 5 min. After igniting the plasma in the chamber a transient effect may uncontrollably influence the sample morphology in the beginning of the deposition process. Therefore, the sample holder is transfered out of the plasma field before igniting the plasma. The plasma ignition is supported by a piezo-element, after flushing the chamber with the precursors for 2 min. The deposition pres-

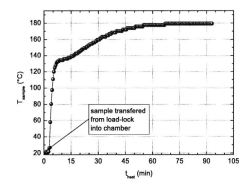

Figure 3.3.: Sample temperature T_{sample} as a function of pre-heating time t_{heat}. The heater temperature T_{heat} is set to 320 °C. During 2-7 min, the sample is transferred via the load-lock into the chamber which is then evacuated. After 60 min under high vacuum, T_{sample} saturates at approximately 180 °C.

sure and the gaseous precursors can be varied via the MFCs depending on the experimental needs.

The samples are radiatively heated from above with actual substrate temperatures (corresponding to T_{rail}) being significantly lower than the heater temperature, T_{heat}. In contrast to the experimental conditions reported in *Grabosch* [88], stating that the sample temperature T_{sample} equals 2/3 of the heater temperature T_{heat}, in the present work T_{sample} is found to be lower. Figure 3.3 illustrates the deposition temperature as a function of pre-heating time of the sample inside the chamber. The sample temperature T_{sample} saturates after 60 min pre-heating. However, to increase the output of samples, a pre-heating time of t_{heat} = 15 min is used to optimize deposition processes (unless stated otherwise). The temperature measured at the rail thermo couple (T_{rail}) corresponds approximately to the sample temperature. T_{sample} is measured via a thermo couple directly attached to the sample surface. The data acquisition is carried out with a *Gemini* logger. The temperature discrepancy measured for the same setup in this work and reported in *Grabosch* [88] arises from the discrepancy of the measurement methods. *Grabosch* measurements

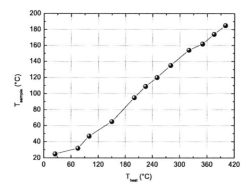

Figure 3.4.: Sample temperature T_{sample} as a function of the heater temperature T_{heat} under high vacuum conditions in the range of $1 \cdot 10^{-8}$ mbar after the sample has been heated up in the chamber for 15 min. For short process times in the range of seconds, one can assume that convection of the precursor gases at standard deposition pressures of 200 - 500 mTorr does not affect T_{sample} or T_{dep} during the deposition.

are done under 1 atmosphere pressure, therefore the air convection leads to a higher T_{sample} detected. In this work the measurements are performed under high vacuum conditions (corresponding to the deposition conditions). Figure 3.4 illustrates the correlation of the sample temperature T_{sample} and the heater temperature T_{heat}. The sample temperature T_{sample} denoted in this work is referred hereafter as the deposition temperature T_{dep}.

Depending on the experimental conditions suitable for a-Si:H/c-Si heterojunction solar cells, the process parameters can be adjusted. The variation of those parameters include:

- **plasma excitation frequency f_{plasma}:** even at the standard deposition frequency of 13.56 MHz it is possible to achieve high-quality a-Si:H layers. However, very-high frequency PECVD with signals of 70 MHz or 110 MHz turned out to be more efficient, and has a favorable impact on (i) the increase of the crystallinity, and (ii) the growth quality of a-SiO$_x$:H material. This detail will be discussed in detail in chapter 6.

- **deposition pressure** p_{dep}: an increase of p_{dep} enables a decrease of ion-bombardment, which is more preferable for the growth conditions in respect of plasma damage. However, if the deposition pressure is too high, the bias voltage increases, supporting the growth of defects in the deposited material. Moreover, high-pressure can support the formation of polysilane powder. Both effects are undesirable. Typical values of p_{dep} are in the range of 200 - 500 mTorr for the PECVD setup used in this work.

- the **gas concentration** χ define the optical and electrical properties, depending on the targeted film properties. In this work, for each experiment the gas concentration ratios will be optimized.

- the **heater temperature** T_{heat}, **and deposition temperature** T_{dep} influence the resulting network of the films drastically: as reported by i.e. *Fujiwara and Kondo* [90], epitaxial (epi) growth of Si occurs in theirs PECVD setup once a critical deposition temperature of $T_{dep} \geq 140$ °C is exceeded. In respect of heterojunction solar cells, several authors confirmed that the interface a-Si:H/c-Si should be abrupt. Hence, epi-growth directly onto the c-Si wafer material would not serve as an abrupt interface.

- **plasma excitation power** P_{rf}: increase of the plasma excitation power influences the deposition rate, as well as a shift of the transition from amorph to micro-crystalline is observed. However, with increasing plasma power, ion-bombardment increases resulting in defective material. Typical values for P_{rf} for the PECVD setup described in this work are in the range of 4 - 20 W (27.9 - 139.7 mW/cm^2).

- **electrode inter-spacing** d_e: the value of d_e, together with the above mentioned deposition parameters, is responsible for the either amorphous or micro-crystalline nature of the deposited layers.

All samples for optical and electrical characterization are either deposited on high-quality silicon wafer material, oxidized wafer substrates or 7059 corning glass substrates, each individually pre-cleaned.

3.1.1.1. Chamber contamination

'Anyone becoming familiar with plasma [..] deposition processes will re-
alize that there are as many deposition reactor designs and experimental
conditions as there are published studies describing plasma deposited [..]
materials.' (*Vossen and Cuomo* [91])

From the statement above, published in 1978, one can assume that it is
not trivial to obtain reliable and reproducible results in respect of plasma
deposited materials. Therefore, it is indispensable for plasma deposition processes to ensure constant deposition conditions. Thus, chamber contaminations due to staining at the chamber walls, or redeposition of doped silicon
material from silicon material deposited onto the sample holder in previous
deposition runs should be avoided. This can be achieved by prior chamber
cleaning or dummy-processes. During the dummy-process a plasma process
with the intended future experiment process parameters is performed without the sample. The sample holder and the chamber walls are covered with
a sufficiently thick layer of the targeted film. For more details on chamber
contamination refer to [92].

3.2. Spectroscopic ellipsometry measurements

Spectroscopic ellipsometry (SE), as a surface and thin film measurement technique, can be used to determine (non-destructive) thin film thicknesses and
thin film optical constants (such as refractive index and extinction coefficient),
surface roughness and optical anisotropy in the near-UV, visible, and near-IR
wavelength ranges. Typically data acquired with the ellipsometer are transmission and reflection intensity, and ellipsometry.

3.2.1. Transmission and reflection

The transmission (T) and reflection (R) measurements acquire the intensity
ratios of T and R, respectively, over a given wavelength region. Therefore, a
beam of light is incident on a sample at some arbitrary angle (θ_i) and partly

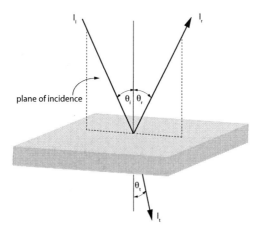

Figure 3.5.: Schematic of the incident, reflected and transmitted beams of light. The angle of incidence θ_i is defined as the angle between the input beam direction and the direction normal to the surface. The plane of incidence is defined as the plane containing the input and output beam, as well as the direction normal to the sample surface.

reflected at angle (θ_r) as well as transmitted at the angle (θ_t) at the boundary of the medium, as illustrated in Fig. 3.5. T and R are defined as the ratio of the light intensity being transmitted (I_t) or reflected (I_r) of the incident light intensity (I_i) [93]. They can be calculated as

$$T = \frac{I_t}{I_i} \qquad \text{and} \qquad R = \frac{I_r}{I_i}. \tag{3.1}$$

3.2.2. Ellipsometry

In general, ellipsometry refers to the measurement of the change in polarization state of light reflected from the surface of a given sample. The measured values are expressed as Ψ and Δ, and derived from the determination of the relative phase change in a beam of reflected polarized light. These values are related to the ratio, ρ, of Fresnel reflection coefficients $\widetilde{R}_p$ and $\widetilde{R}_s$ for p- and

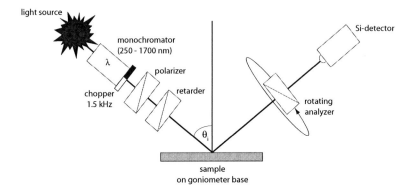

Figure 3.6.: Schematical sketch of a variable angle spectral ellipsometer with rotating analyzer (RAE).

s-polarized[1] light, respectively, *cf*. [93, 94].

$$\rho = \frac{\widetilde{R_p}}{\widetilde{R_s}} = tan(\Psi)e^{i\Delta} \tag{3.2}$$

In this work a variable angle spectral-ellipsometer (VASE) (*J.A. Woolam, INC.*) with a spectral range from 250 nm to 1700 nm is used. The setup consists of a rotating analyzer ellipsometer (RAE) in which the light beam leaves the monochromator, passes through a *fixed* (input) polarizer, is reflected from the sample, passes through a *rotating* polarizer (the analyzer, which is continuously rotating) and then strikes the Si-detector, as illustrated in Fig. 3.6. In general, the detector signal is measured as a function of time, the measured signal is Fourier analyzed in order to obtain the Fourier coefficients a_n and b_n, and finally Ψ and Δ are calculated from a_n and b_n and the known azimuthal angle, P, of the input polarizer. The following equations form the basis of the ellipsometric measurements with a rotating analyzer ellipsometer:

$$tan\Psi = \sqrt{\frac{1 + a_n}{1 - a_n}} \, |tanP|, \tag{3.3}$$

---

[1] The p- and s-directions are the two orthogonal basis directions used to specify the electric field or polarization state of an arbitrarily beam of light.

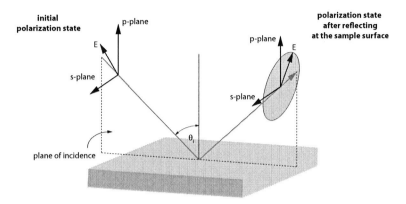

Figure 3.7.: Schematic of an ellipsometric experiment, showing the p- and s-directions. Linearly polarized light is reflected off the sample and detected as elliptically polarized light.

$$cos\Delta = \frac{b_n}{\sqrt{1 - a_n^2}} \cdot \frac{tanP}{|tanP|}.$$

Ellipsometry measurements provides the best results for a film thickness in a wavelength range of the light used for the measurement. Also, roughness features on the sample surface or film interface should be less than ∼ 10 % of the probe beam wavelength for the ellpsometric analysis to be valid. More detailed information about ellipsometer physics and the VASE can be found in Ref. [94].

3.2.3. Modeling and determination of optical constants

Ellipsometry as an optical technique requires an accurate model of the measurement process to analyze the input data. In many cases, the spectral acquisition range and angles of incidence allow the deduction of both thickness and optical constants of the same film. The key components of the ellipsometric models are the optical constants of the substrate and sample layers and the thickness of the layers. The optical constants are parameters which characterize how matter will respond to excitation by an electromagnetic radiation at a given frequency. These optical constants can be expressed as a complex dielectric function by

$$\tilde{\epsilon} = \epsilon_1 + i\epsilon_2, \tag{3.4}$$

or as a complex refractive index

$$\tilde{n} = n + i\kappa. \tag{3.5}$$

The complex index of refraction and the complex dielectric function of a material are related by

$$\tilde{\epsilon} = \tilde{n}^2 \Rightarrow \begin{cases} \epsilon_1 = n^2 - \kappa^2 \\ \epsilon_2 = 2n\kappa \end{cases} \tag{3.6}$$

The real part or index of refraction (n) defines the phase velocity of light in material

$$\upsilon = \frac{c}{n}, \tag{3.7}$$

where υ is the speed of light in the material and c is the speed of light in vacuum. The imaginary part or extinction coefficient (κ) determines how fast the amplitude the wave decreases. The extinction coefficient is directly related to the absorption of a material. Thus, the optical constants expressed by the real and imaginary parts of the complex index of refraction (n and κ) represent the optical properties of a material in terms of how an electromagnetic wave will propagate in that material. Alternatively, the real and imaginary parts of the dielectric function contain the same information in terms of how the material responds to an applied electric field [94].

The absorption coefficient, α, determines how far into a material light of a particular wavelength can penetrate before its absorbed; usually it is equal to the depth at which the energy of the radiation has decreased by the factor of $e^{-\alpha x}$. Its value can be deduced from the extinction coefficient κ after *Davis and Mott* [63] as

$$\alpha = -\frac{1}{\Phi}\frac{d\Phi}{dx} = \frac{2\omega\kappa}{c_0} = \frac{4\pi\kappa}{\lambda}, \tag{3.8}$$

where Φ defines the spectral photon flux density as $\Phi = \int SdF \propto |\tilde{E}|^2$ with the Poynting vector $S = \tilde{E}x\tilde{H}$ and $\tilde{B} = \mu\tilde{H}$ and the surface area F.

A material with low absorption coefficient absorbs light only poorly; if the material is thin enough, it will appear transparent to a certain wavelength. Semiconductor materials have a sharp edge in their absorption coefficient, since light which has energy below the bandgap does not have sufficient energy to raise an electron across the bandgap. Consequently this light is not absorbed [2]. The ratio of incident spectral photon flux density, Φ_i, and absorbed radiation flux density, Φ_a, defines the absorption level, A, for electromagnetic radiation covering a distance x in the given material.

$$A = \frac{\Phi_i}{\Phi_a} = 1 - exp(-\alpha x). \tag{3.9}$$

3.2.3.1. Modeling

To determine the optical bandgap E_G of the amorphous or microcrystalline plasma deposited films in this thesis, the dispersion model for amorphous material based on the absorption edge Tauc formula by *Davis and Mott* [63], and the quantum mechanical Lorentz oscillator model, proposed by *Jellison-Jr. and Modine* [95], has been employed. The SE data – cos (Δ) and tan (Ψ) – are fitted assuming a three layer model (unless stated differently): *surface roughness/a-Si film/corning-glass 7059*, or a four layer model *surface roughness/a-Si film/SiO$_2$/c-Si*. For samples investigated in this work, polished crystalline wafer or corning glass has been used as substrate, as roughness feature on the sample surface or at the film interface should be generally less than 10 % of the probe beam wavelength for the ellipsometric anaylsis to be valid. In case a c-Si substrate is used, the SiO$_2$ layer introduced serves as an optical separation layer. An accurate fit of very-thin films benefits from a sandwiched thermally grown silicon dioxide layer between crystalline wafer and amorphous/microcrystalline plasma deposited layer to enhance the resolution.

3.2.3.2. Determination of the optical bandgap

The optical bandgap is determined from ellipsometry data using mostly the Tauc method [62], whereas a number of different conventions have been established among different authors; and a comparison between those will pinpoint slight differences of E_G. In the following a brief discussion of those conventions for obtaining E_G will be given.

E_{Tauc} The most widely used definition of the optical bandgap, E_g, is based on a formula developed by *Tauc* [62]. The quantity $(\alpha E)^{1/2}$, where α is the experimentally deduced absorption coefficient, is plotted as a function of photon energy $(E = \hbar \cdot \omega)$ and extrapolated using the Tauc equation: at high enough absorption levels $(\alpha > 10^4 \text{ cm}^{-1})$ the absorption constant α has the following frequency dependence

$$[\hbar \cdot \omega \cdot \alpha (\hbar\omega)]^{1/2} = A' (\hbar \cdot \omega - E_{Tauc}), \qquad (3.10)$$

where A' is the density of the localized state constant depending on the material; it can be calculated from the optical density data. The *Tauc* optical bandgaps (E_G) of the deposited films are thereby determined by the intercept of the $(\hbar \cdot \omega \cdot \alpha)^{0.5}$ *versus* $(\hbar \cdot \omega)$ curve in the high absorption region, following the analysis by Davis and Mott [63, 96]. An example of analysis of the optical bandgap using the Tauc-plot is given in Fig. 3.8a.

E_{04} Another approach determines the optical bandgap at a particular absorption level. In particular, with $\alpha = 10^4 \text{ cm}^{-1}$ the optical gap is denoted as E_{04}. The value of the bandgap E_G is distinguished from the absorption edge which corresponds to the minimum energy difference between the lowest minimum of the conduction band and the highest maximum of the valence band. This definition of the optical bandgap has often a large uncertainty in respect of the film thickness estimation; also, the nature of α varying with the energy has not been taken into account, but only its magnitude. An example of the absorption edge method determining E_{04} is given in Fig. 3.8b.

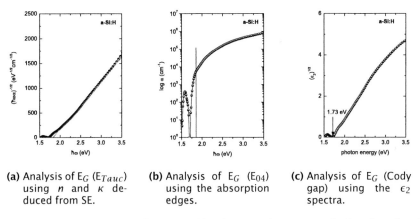

(a) Analysis of E_G (E_{Tauc}) using n and κ deduced from SE.

(b) Analysis of E_G (E_{04}) using the absorption edges.

(c) Analysis of E_G (Cody gap) using the ϵ_2 spectra.

Figure 3.8.: Determination of the optical bandgap E_G of a-Si:H; initial values from fitted SE measurement data of a-Si:H on a SiO_2/c-Si structure.

E_{Cody} The Cody gap, E_{Cody}, is obtained after *Cody et al.* [97] from extrapolation of $(\epsilon_2 E^2)^{1/2}$ data, estimating the energy position of $(\epsilon_2)^{1/2}=0$, *cf.* [98]. Fig. 3.8c illustrates an example of an analysis of the optical bandgap (E_{Cody}) using the ϵ_2 spectra. While Cody reported fundamental inconsistency in the Tauc analysis method for the determination of the optical bandgap in amorphous silicon films, Tauc gaps are generally displayed in this thesis for sake of comparability, as they appear to be more widely reported; hereafter, the systematic bandgap changes observed in this work will be designated as E_G.

3.3. μ-Raman spectroscopy

Micro-Raman (μ-Raman) spectroscopy is used for the observation of changes in the microscopic network of for instance a semiconductor due to growth conditions or other treatment. To understand the Raman-effect, a short overview of Raman theory will be given here. Detailed information can be found in literature in e.g. *Ferraro and Nakamoto* [99], *Suetaka* [100], *Turrell and Corset* [101], *Schrader* [102], *Laserna* [103].

3.3.1. Theoretical aspects of μ-Raman spectroscopy

When monochromatic radiation impinges upon a molecule of a semiconductor, it may be reflected, absorbed or scattered in some manner. Light scattered from a molecule has several components - the Rayleigh scatter, and the Stokes and Anti-Stokes Raman scatter (approximately only 1×10^{-7} of the scattered light is Raman scatter). *Rayleigh scattering* designates a process without any change of frequency. *Raman scattering* specifies a change in frequency or wavelength of the incident radiation. The spontaneous Raman effect occurs, when the incident photons with the energy of $h\nu_0$ thus interact with that molecule and excite an electron from the ground state to a virtual energy state. Relaxing into a vibrational excited state generates *Stokes Raman scattering*. The Raman shifted photons can be either of higher or lower energy, depending upon the vibrational state of that molecule. Since there are a small number of molecules, which are existent in a elevated vibrational energy level, the scattered photon can actually be scattered at a higher energy. In this case, the Raman scattering is then called *Anti-Stokes Raman scattering*. It is the amount of energy change by that photon (either lost or gained), which is characteristic of the nature of each bond (vibration) present and provides the chemical and structural information.

Not all vibrations will be observable with Raman spectroscopy (depending upon the symmetry of the molecule). Hence, the amount of energy shift for a C-H bond is different to that seen with a Si-H bond. By analyzing all those various wavelengths of scattered light, it is possible to detect a range of wavelengths associated with the different bonds and vibrations.

3.3.2. The Raman spectroscope

The Raman spectroscope used in this work developed by *JOBIN Labram* consists of a microscope confocally coupled to a 300 mm focal length spectrometer. The excitation light, supplied by an Ar^+ ion laser with a wavelength of 488 nm, is guided and focused on the sample through a microscope, as well is the backscattered Raman signal. The excited volume probed in the Raman analysis depends on the absorption coefficient, α, of the sample at the excitation

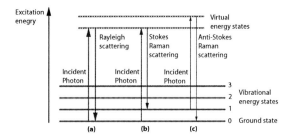

Figure 3.9.: Simplified energy level diagram illustrating Raman scattering: (from left) Rayleigh scattering, Stokes Raman scattering and anti-Stokes Raman scattering.

wavelength used. The penetration depth ($= 1/\alpha$) for the Ar^+ ion laser used for this experiments is in the range of 50 nm or less for amorphous silicon and about 500 nm or less for crystalline silicon. The output power of the laser can be varied from 15 to 45 mW. Several filters decrease this power output to 20 % or less. Espacially for measurements of thin amorphous silicon layers, the weak power of the laser avoids any laser induced crystallization. The resolution of the recorded spectra lies at about 1 cm^{-1}, the spectra are recorded at room-temperature in this work, unless stated differently.

3.3.3. Data analysis

The crystalline volume fraction measured via Raman spectroscopy is a criterion to describe the silicon materials in its transition zone from amorphous to crystalline. In Raman spectroscopy the standard peaks for crystalline and amorphous silicon appear at 520 cm^{-1} and 480 cm^{-1} wavenumbers, respectively. Therefore, the position of the Raman peaks for micro-crystalline silicon would determine prevalence of either crystalline or amorphous structure in the deposited films. Under this considerations, the peak at 500 cm^{-1} would indicate dominance of crystalline fraction in the examined samples, whereas the position of the peak close to 480 cm^{-1} attributes amorphous character of micro-crystalline film. The terminology of micro-crystalline is not accurately defined in literature, as one can use the term 'micro-crystalline layer' or 'close to the transition to crystallinity' for a PECV deposited layer, which contains only a fraction of micro-crystals. Hereafter, all μc-Si:H films or a-Si:H layers with a micro-crystalline fraction are referred to as 'micro-crystalline' films, whereas 'crystallinity' refers to the degree of structural order in silicon (usually specified as a volume percentage that is crystalline).

3.4. Conductivity

The measurement of the dark- and photo-conductivity allows one to evaluate the basic electronic properties (such as doping concentration, and carrier mobility) of PECV deposited a-Si:H layers and its alloys. Therefore, a brief description of the experimental methods applied to obtain dark- and photo-conductivity is given below.

3.4.1. Dark- and photo-conductivity of PECV deposited layers

The dark conductivity (σ_{dark}) and photo-conductivity (σ_{ph}) of the deposited a-Si:H films and its counterparts are determined using the standard four-point-probe method as well as the *transfer length method* (TLM) method. In the four-point-probe method a current passes through two outer probes, whereas the generated voltage is measured only through the inner probes, which allows the measurement of the substrate resistivity. Detailed information can be found elsewhere [104]. The TLM is based on a measurement technique originally proposed by *Shockely*; the method consists of current-voltage characteristic measurements, using a coplanar electrode configuration as illustrated in Fig. 3.10 schematically. The implementation of this method implies unequally spaced contacts (as displayed in Fig. 3.10). The voltage is measured between the contacts to ensure there is only bare semiconductor between the contacts; no other contacts should interfere with the measurement.

The total resistance between any two contacts can be then expressed (derivation can be found in [104]) as

$$R_T = \frac{\rho_s d}{Z} + 2R_c \approx \frac{\rho_s}{Z} \left(d + 2L_T \right). \tag{3.11}$$

The plot of the total resistance as a function of the contact spacing d can

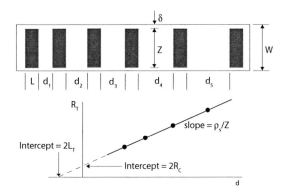

Figure 3.10.: *Transfer length method* (TLM) test structure and a plot of total resistance as a function od contact spacing, *d*. After [104].

be derived once the values of R_T are measured for various contact spacings. From this plot, is is possible to extract (i) the *sheet resistance* by the slope $\Delta(R_T)/\Delta(d) = \rho_s/Z$, (ii) the *contact resistance* by $R_T = 2R_c$ at the intercept at $d = 0$, and (iii) the *specific contact resistivity* with ρ_s known from the slope of the plot by $-d = 2L_T$ at the intercept at $R_T = 0$.

An electrometer (*Keithley (6517)*) or *Hewlett Packard 4156A* semiconductor parameter analyzer is used to measure the resistivity in the dark or under one-sun illumination at standard test conditions at 25 °C.

3.4.2. Activation energy

The activation energy E_T is typically defined as E_C-E_F for as *n*-type wafer or E_F-E_V for *p*-type wafer. E_T can be obtained by the Arrhenius plot obtained from dark conductivity measurements, whereas σ_d and E_T are related by

$$\sigma_d = \sigma_0 exp\left(-\frac{E_T}{k_bT}\right),\tag{3.12}$$

where σ_0 denoted the microscopic prefactor ($150\ \Omega^{-1}cm^{-1}$ after [105]). The measurement is carried out using either an evacuated high-temperature chamber (with a temperature ranging from 300 K to 550 K) or optionally with an evacuated low-temperature chamber (*Cryophysics*, with a temperature ranging from 30 K to 320 K). Electric contacting is established via four tungsten probes driven by a source measure unit (SMU) 236/237 from *Keithley*.

3.5. Solar cell characterization

One of the most important characterization methods for fabricated solar cells are the current-voltage characteristics (under illumination and in the dark), revealing parameters such as V_{oc}, J_{sc}, R_s, R_p, FF and η. Quantum efficiency measurements provide an analysis of wavelength dependent cell characteristics. Therefore, the setups used in order to obtain the parameters mentioned above are described in the following.

3.5.1. I-V characteristics

The most fundamental of solar cell characterization techniques is the measurement of the cell efficiency. Standardized testing allows the comparison of devices manufactured at different companies and laboratories using various technologies.

3.5.1.1. Light I-V

The *Standard Test Condition* (STC) for solar cell characterization are

- Air mass 1.5g spectrum (AM1.5g) for terrestrial cells

- Intensity of 1000 W/m^2 (100 mW/cm^2, one-sun of illumination)

- Cell temperature of 25 °C

- Four-point probe technique to remove the effect of probe/cell contact resistance

The current-voltage (I-V) characteristics have been carried out using a custom-built tester described elsewhere [106]. The STC for solar cells described above have been strictly adhered to in the framework of this thesis. For example, a deviation in temperature introduces errors in the values of V_{oc}. One-sun illumination is quite intense; therefore the cell is placed on a commercially available, water-cooled brass block. A thermocouple is inserted in the block and the control system is adjusted so that the required temperature of 25 °C is kept constant. To contact the cell using the four-point-probe technique, a current and voltage probe on top of the cell and a current and voltage probe on the bottom of the cell are used. The metal block act as the rear (current and voltage) contact. The top contacts are usually paired since it is insufficient to have a single voltage and current probe. Ideally the probes make good contact with the cell and the voltage and current probes are within close proximity but not touching each other [2].

3.5.1.2. Dark I-V

Dark I-V measurements provide direct diode properties of a solar cell. From the slope of the dark I-V plot at 0 V the parallel resistance R_p can be extrapolated as

$$R_p = \frac{dV}{dI}\bigg|_{V=0[V]}. \tag{3.13}$$

A linear graph of current *vs.* voltage reveals very little information about the diode, much more information is revealed from a semi-log plot. A typical dark current-voltage characteristic for a solar cell is plotted in semi-log scale displayed in Fig. 3.11.

In addition, dark I-V characteristics allow to draw conclusions from the idealitiy factor n_{id1} and n_{id2}. Any non-ideality in the fill-factor is assigned to low a R_p or high n_{id2}; a higher n_{id1} will lead to low voltages. This tendency could be due to a low bulk lifetime in the substrate, or poor surface passivation quality. A higher n_{id2} will also lead to lower voltages, especially in the voltage range typical for the maximum power point. Usually this is due to bulk lifetime effects, but it can also be a complex combination of shunting effects and an injection-level-dependent lifetime. A low shunt resistance will reduce the fill factor and maximum power point voltage, but it will have negligible effect on the open-circuit voltage at the intensity of one-sun, *cf.* [107]. The series

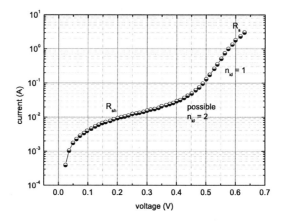

Figure 3.11.: Dark current-voltage characteristic of a silicon solar cell, after [2].

resistance contributions can be extrapolated by

$$R_{s,dark} = \frac{V_{dark}(I_{sc} - V_{oc}(I_{sc}))}{I_{sc}},$$ (3.14)

and

$$R_{s,light} = \frac{V_{oc}(I_{sc} - I_{mp}) - V_{light}(I_{sc} - I_{mp})}{I_{sc} - I_{mp}}.$$ (3.15)

3.5.2. Spectral response (SR)

While the quantum efficiency specifies the electron output of the cell compared to the incident photons, the spectral response (SR) is defined as the ratio of the current generated by the cell to the incident power on the cell surface. The I_{sc} is measured under monochromatic illumination with the intensity $\Phi_0(\lambda)$. SR is then defined as

$$SR(\lambda) = \frac{I_{sc}(\lambda)}{\Phi_0(\lambda)}.$$ (3.16)

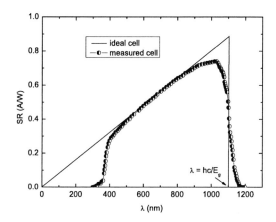

Figure 3.12.: SR of a silicon solar cell under glass. At short wavelengths (below 400 nm) the glass absorbs most of the light and the cell response is very low. At intermediate wavelengths the cell approaches the ideal response. At long wavelengths the response fall back to zero. Silicon is an indirect band gap semiconductor so there is not a sharp cut off at the wavelength corresponding to the band gap ($E_G = 1.12$ eV). After [2].

Usually the spectral response is measured in units of [A/W]. A typical SR curve is illustrated in Fig. 3.12. The SR is obtained using a custom-build setup as illustrated in Fig. 3.13

3.5.2.1. Quantum efficiency

The quantum efficiency (QE) can be calculated based on the SR measurements: the quantum efficiency can be determined from the spectral response by replacing the power of the light at a particular wavelength with the photon flux for that wavelength. The QE can be defined as the ratio of the number of carriers collected by the cell to the number of photons of the incident light:

$$SR(\lambda) = \frac{q\lambda}{hc}QE(\lambda). \tag{3.17}$$

A solar cell under illumination at short circuit generates a photocurrent

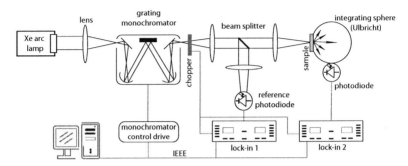

Figure 3.13.: Schematic illustration of setup for transmission measurements. The same setup is used for reflection and spectral response measurements; therefore, the arrangement of the sample can be re-adjusted depending on the intended measurement.

depending on the incident light. The photocurrent density (J_{sc}) can be related to the incident spectrum via the cell's quantum efficiency (QE). Therefore the QE is a key quantity to describe the solar cell performance under different conditions [31].

$$J_{sc} = q \int \Phi_s(E) \cdot EQE(E) dE = \frac{q}{hc_0} \int \lambda \cdot EQE(\lambda) \cdot \Phi_s(\lambda) d\lambda, \qquad (3.18)$$

where $\Phi_s(E)$ indicates the incident spectral photon flux density, q is the electronic charge, c_0 is the speed of light in vacuum and h is the Planck constant.

The external quantum efficiency (EQE) does not depend on the incident spectrum, so to say light that does not enter the cell, or light that leaves the cell again after entering, does not contribute to the photocurrent. Thus, the EQE depends upon the absorption coefficient of the solar cell material (expressed by the probability, f_{abs}, that an incident photon, carrying the energy E, will be absorbed), and the efficiency of charge separation and charge collection in the device (enunciated by the probability f_c). A EQE curve for an ideal solar cell is shown in Fig. 3.14.

The EQE of a solar cell includes the effect of optical losses such as transmission and reflection. A quantity that depends less strongly on the optical design (or a quantum efficiency of the light left after the reflected and transmitted light has been lost) can be given by the internal quantum efficiency (IQE). The

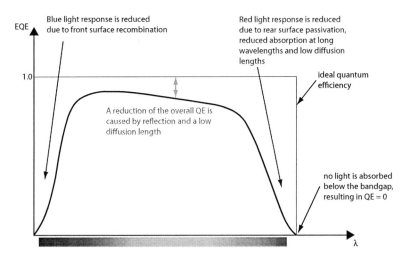

Figure 3.14.: The quantum efficiency of a silicon solar cell. Quantum efficiency is usually not measured below 350 nm as the power from the AM1.5 contained in such low wavelengths is low. After [2].

IQE is taking the hemispherical reflectance data, R, of the device into account to obtain a corrected EQE curve. The IQE gives hereby the probability that an incident photon of a certain wavelength λ (carrying the energy E [2]), which is neither reflected from the device surface nor transmitted through the device, delivers one electron to the external circuit (J_{sc}). The IQE is then defined as

$$IQE = \frac{EQE}{(1-R)} = \frac{f_{abs}f_c}{(1-R)}.$$
(3.19)

The denominator (1-R) determines the amount of incident photons in respect of the reflectance R.

3.5.2.2. IQE analysis - effective diffusion length

The effective diffusion length (L_{eff}) of the minority carriers in the bulk can be extracted from the intercept of the inverse measured IQE characteristic

---

[2]A convenient rule for converting between photon energies, in electron-Volts, and wavelengths, in nm, is E/eV = 1240/(λ/nm).

(IQE^{-1}) *versus* the absorption length of the light, L_α, *cf.* [108]. For the case that $\alpha^{-1} < W$,

$$IQE^{-1} = 1 + \frac{cos\Theta}{L_{eff}}\alpha^{-1}, \qquad (3.20)$$

where Θ defines the angle of the incident light through the solar cell (depending on a potential surface-texturing). L_{eff} is defined based on the exact knowledge of the absorption and dispersion of the light in a cell, and can be expressed as

$$L_{eff} = L_b \frac{D_{n,p} + S_{eff}L_b tanh\left(\frac{W}{L_b}\right)}{S_{eff}L_b + D_{n,p}tanh\left(\frac{W}{L_b}\right)}, \qquad (3.21)$$

where L_b describes the bulk diffusion length, W is the thickness of the bulk, and $D_{n,p}$ define the diffusion constants for minority carriers in the bulk. The dependence of IQE^{-1} as a function of α^{-1} provides a linear characteristic (*cf.* Eq. 3.20) with the slope of cos Θ/L_{eff}. Thus, with Θ known, L_{eff} can be extrapolated. Once S_{eff} and L_b are know values, the backward saturation current density, J_{0b}, can be determined after Eq. 3.21 as

$$J_{0b} = \frac{qDn_i^2}{N_B L_{eff}}. \qquad (3.22)$$

3.5.3. Suns-V$_{oc}$

The Suns-V$_{oc}$ technique allows to deduce the open-circuit voltage as a function of light intensity. The material quality as well as the shunting in the cell can be assessed by analyzing the Suns-V$_{oc}$ curve. The measurement itself is very similar to the I-V measurements except that Suns-V$_{oc}$ uses a second (separate) solar cell to monitor the illumination intensity, whereas I-V uses the J$_{sc}$ of the solar cell. Probing either the silicon p$^+$ or n$^+$ regions directly or probing the metalization layer (if present), the illumination-V$_{oc}$ curve is measured at the open-circuit voltage, so its free from the effects of series resistance. Comparing the Suns-V$_{oc}$ curve to the final I-V curve deduced from solar simulator gives a precise measurement of the series resistance in the cell. The Suns-V$_{oc}$ stage is ideal for monitoring of single process steps (e.g. before and after formation of TCO and metal contacts). This allows the optimization and monitoring of steps to maintain V_{oc}, obtain good ohmic contacts and avoid shunting.

3.5.3.1. Suns-V_{oc} setup

A Xenon flash lamp with neutral density filters providing a slowly decaying light is used for illumination. A temperature controlled brass-chuck wafer stage with a build-in monitor solar cell is set to a temperature of 25 °C, as the V_{oc} during measurement strongly depends on the measurement temperature. It is possible to measure heterojunction solar cells without any front or back metalization, if both the *n*- and *p-type* regions can be probed. The position (height) of the flash lamp above the stage is adjustable to fine-tune the light intensity range. The range of illumination intensities falling on the sample is very important in allowing the most value to be extracted from a measurement. The intensity range is chosen with the use of neutral density filters (which are placed between flash lamp and sample), combined with the stand height. The intensity should range from greater than 1 sun down to less than 0.01 suns, which allows the plotting of an I-V curve that includes V_{mp} and V_{oc} under one-sun conditions.

3.5.3.2. Data analysis

Figure 3.15 illustrates an exemplary Suns-V_{oc} curve and the corresponding I-V curve obtained by a solar simulator. Unless the J_{sc} is known with precision from solar simulator measurements, the J_{sc}, the J_{mp} and efficiency values are as uncertain as J_{sc}. However, the spreadsheet provides the V_{mp}, V_{oc}, pseudo FF and shunt resistance, which are very insensitive to the uncertainties in J_{sc}. Therefore, when using the Suns-V_{oc} it is recommended to track these parameters as indications of device quality and process control. One of the most relevant parameter is the V_{oc} at 0.1 suns, rather than at 1 sun: the distribution of V_{oc} at 0.1 suns is very sensitive and noise free indication of the cell quality in the relevant range of operation, whereas the V_{oc} at 1 sun is taken very near the actual operating point of the solar cell at V_{mp}. Therefore, the V_{oc}

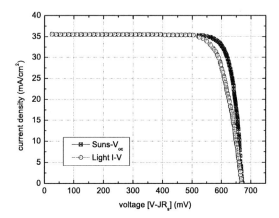

Figure 3.15.: Exemplary Suns-V_{oc} curve with no impact of series resistance and corresponding I-V curve measured by a solar simulator. Both the material quality and the shunting are nicely displayed. After [107].

at 0.1 suns is sensitive on the material quality (when the cell is at V_{mp}), as well as the effect of the shunting on the solar cell performance. More information about Suns-V_{oc} measurements can be found in [107].

A double-diode expression can be fitted, providing information about the parameters J_{01} and J_{02} (which determine the recombination currents associated with an ideality of $n_{id1} = 1$ and an ideality of $n_{id2} = 2$). The impact of those parameters is illustrated in Fig. 3.11, in section 3.5.1.2.

3.6. Effective carrier lifetime

A measured lifetime obtained by quasi-steady state photoconductance (QSSPC) or transient photoconductance (TPC) conditions (refer to sections 3.6.1 and 3.6.2) can be referred as *effective lifetime*, τ_{eff}. Within this work both methods are used, the QSSPC, which allows to obtain τ_{eff} by a injection depended light flash, as well as the microwave photoconductance decay (μ-WPCD) method, which enables to extrapolate τ_{eff} by the detection of reflected

microwaves. The effective lifetime quantifies the rate of recombination occurring within a solar cell; it represents the combined effect of all competing
recombination paths, and can be expressed by the following equation

$$\frac{1}{\tau_{eff}} = \frac{1}{\tau_{radiative}} + \frac{1}{\tau_{Auger}} + \frac{1}{\tau_{SRH}} + \frac{1}{\tau_{surface}} + \frac{1}{\tau_{emitter}} \tag{3.23}$$

Radiative, Auger and SRH recombination may occur in general in the bulk of
a crystalline wafer, and can be therefore combined to a bulk recombination
rate, U_{bulk}; and analogue to a bulk recombination lifetime, τ_{bulk}:

$$\frac{1}{\tau_{bulk}} = \frac{1}{\tau_{radiative}} + \frac{1}{\tau_{Auger}} + \frac{1}{\tau_{SRH}} \tag{3.24}$$

The effective recombination rate, U_{eff}, can be defined as the sum of all individual recombination rates, discussed in section 2.1:

$$U_{eff} = U_{bulk} + U_{surface} + U_{emitter} \tag{3.25}$$

Analog to *Sinton* [107], interpreting the effective lifetime depends on the
kind of sample: the impact of all recombination paths depend on the excess carrier density, Δn, because the distinctive recombination path in a given
sample may vary depending on the injection level (Δn). That means different recombination parameters might be extracted at different injection levels.
Samples which are of interest in this thesis are (i) silicon wafers either passivated symmetrical by an dielectric layer e.g., silicon oxide or amorphous silicon and (ii) silicon wafers with plasma deposited high-low junctions or plasma
deposited doped regions (emitter).

- *sample structure (i)*

  Fig. 3.16 illustrates such a sample configuration. Equation 3.25 can be
  reduced and the measured effective lifetime is then a combination of
  bulk and surface recombination:

$$U_{eff} = U_{bulk} + 2U_{surface} \Rightarrow \frac{\Delta n}{\tau_{eff}} = \frac{\Delta n}{\tau_{bulk}} + \frac{S_{front}\Delta n_s}{W} + \frac{S_{back}\Delta n_s}{W} \tag{3.26}$$

  When the bulk lifetime is sufficiently long to allow generated carriers to

Figure 3.16.: Sample configuration for lifetime measurements.

reach both surfaces, and when S is sufficiently low, $\Delta n = \Delta n_s$, resulting in

$$\frac{1}{\tau_{eff}} = \frac{1}{\tau_{bulk}} + \frac{S_{front}}{W} + \frac{S_{back}}{W} \tag{3.27}$$

where $S_{front} = S_{back} = S$ holds, when the dielectric layer is deposited symmetrical on each side of the silicon wafer.

- *sample structure (ii)* For samples with highly doped regions, such as an emitter or BSF on a silicon wafer, equation 3.25 can be reduced and the measured effective lifetime is then a combination of bulk and surface recombination:

$$U_{eff} = U_{bulk} + 2U_{emitter/BSF}$$
$$\Rightarrow \frac{\Delta n}{\tau_{eff}} = \frac{\Delta n}{\tau_{bulk}} + \frac{J_{0,front}np}{qWn_i^2} + \frac{J_{0,back}np}{qWn_i^2}, \tag{3.28}$$

where n_i is the intrinsic carrier concentration in silicon. This expression can be further simplified to

$$\frac{1}{\tau_{eff}} = \frac{1}{\tau_{bulk}} + \frac{J_{0,front}\left(N_{dop} + \Delta n\right)}{qn_i^2W} + \frac{J_{0,back}\left(N_{dop} + \Delta n\right)}{qn_i^2W}, \tag{3.29}$$

where N_{dop} the background dopant density and J_{front}, J_{back} the emitter or BSF saturation density. Although n_i in equation 3.29 is strongly temperature dependent, the ratio J_0/n_i^2 is rather not [107].

3.6.1. Microwave photoconductance decay

Investigations on PECV deposited a-Si:H films and its alloys have been carried out measuring the carrier lifetime by the laser-induced microwave-detected photo conductance decay (μ-WPCD) method, cf. [109, 110]. A microwave relaxometer (MWR-1SI-2A) from *TELECOM-STV* is used. The basis of this measurement method is charge carrier injection in semiconductor by irradiating a spot like area with a laser pulse and determination of charge carrier concentration versus time. The lifetime had been obtained by mapping the lifetime values on a circular area of 1 cm diameter in 1 mm steps and calculating the average value. The main advantage of the μ-WPCD method compared to the QSSPC method (described in section 3.6.2) is the very small expansion of the light pulse; making measurements with high spatial resolution possible. The wavelength of irradiation is less than that of the absorption edge. The charge carrier concentration is estimated by microwave reflection. The relaxation constant is derived from non-equilibrium charge carrier bulk-lifetime. After excitation the excess carrier density can be expressed as

$$\Delta n(t) = \Delta n_0 e^{-\frac{t}{\tau_{eff}}}, \tag{3.30}$$

whereas Δn_0 denotes the excess carrier density at the time t = 0. The transient decay of the photo-conductivity $\Delta\Xi(t)$ correlated with the decay of Δn as

$$\Xi(t) = q\left(\mu_n \Delta n(t) + \mu_p \Delta p(t)\right) W, \tag{3.31}$$

where $\mu_{n,p}$ are the electron and hole mobilities, respectively. For μ-WPCD measurements, the photo conductivity is detected by the reflection of microwaves. The simplified formula for a homogeneous, defect-free wafer with a spatially uniform carrier lifetime is given by

$$\frac{1}{\tau_{eff}} = \frac{1}{\tau_{bulk}} + \frac{1}{\tau_{surface} + \tau_{diff}} \tag{3.32}$$

where τ_{eff} is the effective lifetime (measured), τ_{bulk} is the bulk lifetime, $\tau_{surface}$ is the characteristic surface recombination lifetime component determined by wafer thickness (W), the surface recombination velocity (S_{eff}) (corre-

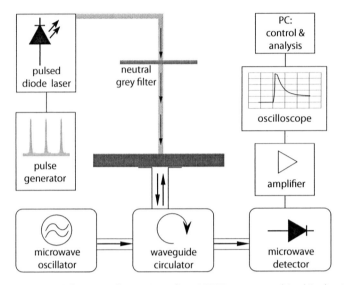

Figure 3.17.: Schematic illustration of a μ-WPCD setup used in this thesis.

lated as $\tau_{surface} = W/2 \cdot S_{eff}$), and the τ_{diff} is the characteristic time of carrier diffusion from the wafer layer center to its surface (given by $\tau_{diff} = W^2/\pi \cdot D_{n,p}$, where $D_{n,p}$ is the ambipolar diffusity, with $D_n = 28$ cm^2/s for p-type wafer and $D_p = 11$ cm^2/s for n-type wafer), cf. [111]. Fig. 3.17 illustrates a schematic of the μ-WPCD setup used.

3.6.2. Quasi steady state and transient photoconductance decay

The overall recombination is evaluated by measuring the excess carrier density dependent effective lifetime with a Sinton Consulting WCT-120 system providing both the transient photoconductance technique (TPCD) and the quasi-steady-state photoconductance technique (QSSPC). A detailed description of the QSSPC and the TPCD method and the setup used in this work can be found in Ref. [112, 113]. After *Nagel et al.* [114], the carrier density can be expressed as

$$\frac{d\Delta n(t)}{dt} = G(t) - U(t) + \frac{1}{q}\Delta J. \tag{3.33}$$

For the case of a homogeneous distributed carrier density, or homogeneous generation rate G, the last term in Eq. 3.33 is negligible. Bearing in mind that $U_{eff} = \Delta n / \tau_{eff}$, the effective lifetime can be written as

$$\tau_{eff} = \frac{\Delta n(t)}{G - \frac{d\Delta n(t)}{dt}}. \tag{3.34}$$

This expression reduces to the transient expression in case G = 0, and to the QSS case when $d\Delta n/dt = 0$.

QSSPC   As for the QSSPC method, the carrier lifetime can be obtained via the photoconductance based on the following method. The silicon wafer sample is exposed to a long, exponentially decaying light pulse. If the excess carrier populations are close to be in 'steady-state', then the generation and recombination rates are in balance. The excess carrier density, Δn, can be achieved by measuring the sheet conductivity as a function of time inductively coupled to the sample by an underlaying rf-coil. Then each moment in time corresponds to different injection levels. The measured sheet conductivity can be then used to calculate the excess carrier density (or Δn) using existing models for carrier mobilities. The inductively measured excess photoconductance is given by

$$\sigma_L = q \left(\Delta n_{av} \mu_n + \Delta p_{av} \mu_p \right) \tag{3.35}$$

where $\Delta n_{av} = \Delta p_{av}$ is the average excess carrier density, μ_n and μ_p are the electron and hole mobilities, and W is the wafer thickness. The carrier mobilities depend on doping type, concentration and injection level. The models used were established by *Dannhaeuser* [115], *Krausse* [116] and [52] and appear to be consistent with the present day theory and experiment. The intensity of the flash during measurement is taken as a function of time and converted into a generation rate, G, of electron-hole pairs in the sample. This requires the knowledge of an optical constant, an estimate of the amount of incident light that is absorbed in the sample though, which is mainly based on optical models on know values of the absorption coefficients and refractive indices of a silicon wafer and various surface films. *Sinton* [107] gives a detailed description about possibilities obtaining the optical constant, i.e., using the software PC1D [117]. The effective lifetime can be calculated via a steady-state condition $\Delta n = G \cdot \tau_{eff}$.

TPCD   The transient PCD method does not require a value for the optical constant, but contrary to the QSSPC method, this technique is only appropriate for the evaluation of photogenerated carrier lifetimes appreciably greater than the flash turn-off time; or when there is a significant excess carrier density in the sample after the flash has terminated. In the TPCD experiment the sample is subjected to a very short and fast pulse of light (in a time frame of typically 10-20 μs), which decays back rapidly. After the flash finished, and the carriers have re-distributed evenly across the wafer, the slower decay of sheet conductivity can be observed as a function of time and further converted into the average excess carrier density Δn. The effective carrier lifetime at each excess carrier density is determined via $\tau_{eff} = -\Delta n/(d\Delta n/dt)$. Figure 3.18 illustrates a schematic of the used setup for QSSPC and TPCD measurements.

Measurement mode   Evaluating the measured effective lifetime over the widest possible range of the excess carrier density (ECD), measurements of QSSPC/TPCD data analyzed in the corresponding QSS, generalized, or transient mode, are combined. An example is displayed in Fig. 3.19. After *Kerr and Cuevas*, the strength in using both QSSPC and TPCD methods to determine the effective carrier lifetime is that while the two methods are complementary, they also have different dependencies on the system calibration constants. The TPCD method is quite independent of the calibration constants for a system with linear response, while the QSSPC method relies on them. As can be seen in Fig. 3.19, the excellent overlap in the lifetime measurements as a function of injection level for the two methods is therefore a good verification that the calibration constants were accurately determined.

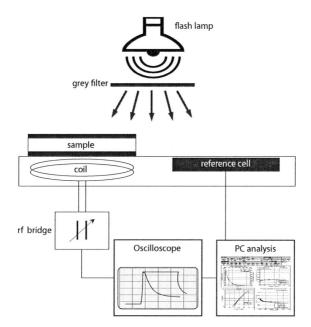

Figure 3.18.: Schematic of the inductively coupled photoconductance apparatus used for the effective lifetime measurements.

3.6.3. Device characteristics determined from effective lifetime

The measured values of τ_{eff} can be used to obtain the effective surface recombination velocity S_{eff}, the implied V_{oc}, or the emitter saturation carrier density J_{0E}. For extrapolation of those for the case of a homogeneous silicon wafer, a derivation for the photo-generated excess carrier density is necessary. Here, only a brief discussion of this process for the case of a p-type wafer is presented. More detailed information of device characteristics determination can be found elsewhere [28, 118].

3.6.3.1. Surface recombination velocity (SRV)

For calculation of the surface recombination velocity, the ambipolar transport equation for the one-dimensional case under a low-injection condition can be

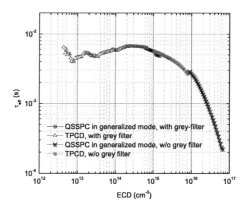

Figure 3.19.: Lifetime measurements taken at different modes and superimposed to evaluate τ_{eff} over the widest possible range of excess carrier density (ECD). This procedure leads to minor discontinuities in the measurement data.

expressed by

i.e., for p-type wafer:
$$\frac{\partial \Delta n}{\partial t} = G - \frac{\Delta n}{\tau_{bulk}} + D_n \frac{\partial^2 \Delta n}{\partial x^2} \qquad (3.36)$$

A general solution for this equation for $t > 0$ ($G = 0$) in equilibrium is found after *Kousik et al.* [119] as

$$\Delta n(x,t) = e^{-t/\tau_{bulk}} \sum_{k=0}^{\infty} \left(A_k \cos(\alpha_k x) + B \sin(\alpha_k x) \right) e^{-\alpha_k^2 D_n t} \qquad (3.37)$$

with

$$\frac{1}{\tau_k} = \frac{1}{\tau_{bulk}} + \alpha_k^2 D_n, \qquad (3.38)$$

and

$$\tan(\alpha_k W) = \frac{D_n \alpha_k (S_1 + S_2)}{D_n^2 alpha_k^2 - S_1 S_2}, \qquad (3.39)$$

whereas A_k, B_k and α_k denote determining factors. From Eq. 3.37 it is evident that $\Delta n (x,t)$ consists of an infinity number of modes which are independently exponentially decaying with τ_k. For sufficient high t, only the base mode with τ_0 is ascertains the decay. τ_0 is therefore assigned as τ_{eff}. From Eq. 3.32 can be concluded, that a better passivation leads to a higher τ_{eff}. For calculation of S_{eff} one can assume that both surfaces provide a sufficiently low recombination velocity and have the same values ($S_{eff} = S_{front} = S_{back}$), as the sample structure is symmetric. Thus, after *Luke and Cheng* [120], Eq. 3.39 can be simplified to

$$\tan \frac{\alpha_0 W}{2} = \frac{S}{D_n \alpha_0}. \tag{3.40}$$

Eq. 3.38 and Eq. 3.40 can be combined to

$$S = \sqrt{D_n \left(\frac{1}{\tau_{eff}} - \frac{1}{\tau_{bulk}} \right)} \tan \left(\frac{W}{2} \sqrt{\frac{1}{D_n} \frac{1}{\frac{1}{\tau_{eff}} - \frac{1}{\tau_{bulk}}}} \right) \tag{3.41}$$

Hence, for good passivated surfaces, τ_{eff} is independent from D_n, but only depending from the recombination in the bulk and at the top- and bottom surface. Though, the main error in determining S_{eff} arises from the uncertainty in τ_{bulk}. Schmidt and Aberle [121] reported the lowest S_{eff} with 5.5 cm/s for a 1000 Ωcm n-type wafer, passivated with a PECVD SiN_x film. However, the uncertainty of S_{eff} could have been as high as 10 cm/s [27] depending on the value used for τ_{bulk}. Therefore, it is useful to calculate S_{eff} for the case that no Shockley-Read-Hall recombination is considered ($\tau_{bulk} \rightarrow \infty$). For a sufficient small surface recombination velocity, Eq. 3.41 can be simplified to

$$\frac{1}{\tau_{eff}} = \frac{1}{\tau_{bulk}} + \frac{2S}{W} \tag{3.42}$$

3.6.3.2. Implied V_{oc}

The implied open-circuit voltage can be obtained by the traditional open-circuit voltage decay (OCVD) technique, where the V_{oc} is measured after the illumination has been terminated, or by the quasi-steady-state V_{oc} (QssV_{oc})

technique, where the V_{oc} is measured during illumination under a photographic flash-lamp. Hereby, the implied (expected) open-circuit voltage can be deduced from the separation of the quasi-Fermi-levels by lifetime measurements and is then calculated via the expression

$$V_{oc} = \frac{kT}{q} ln \left(\frac{np}{n_i^2} \right)$$
(3.43)

where n and p are the total electron and hole concentrations. For p-type silicon e.g., $p = N_A + \Delta n$ and $n = \Delta n$. From Eq. 3.43 follows that the calculation of V_{oc} is quite sensitive to the wafer resistivity and wafer type [107]. These methods are described in more detail in [122] and [38].

3.6.3.3. Emitter current saturation density J_{0E}

In case the sample exhibits identical doping profiles on both surfaces, the emitter current saturation density J_{0E} can be obtained from the calculation of S_{eff}. For low-injection conditions, J_{0E} can be expressed by

$$J_{0E} = \frac{q n_i^2}{N_A} S_{eff}.$$
(3.44)

CHAPTER 4.

APPLICATION OF WIDE-GAP AMORPHOUS SILICON CARBIDE FILMS (A-SI$_X$C$_{1-X}$:H$_Y$) TO HETEROJUNCTION SOLAR CELLS FOR USE AS EMITTER

4.1. Introduction

Hydrogenated amorphous silicon (a-Si$_x$:H$_y$) has been accepted as a suitable heterojunction material for a-Si:H/c-Si solar cells, and conversion efficiencies exceeding 21 % have been reported recently [22]. Nevertheless, photon absorption in the emitter of a heterojunction solar cell leads to a considerable current loss with increasing a-Si:H layer thickness due to the high recombination in this layer [60, 123]. To improve J$_{sc}$ in a-Si:H/c-Si solar cells further, it is preferable to employ an a-Si:H-based alloy that has larger optical bandgap than a-Si:H to suppress light absorption in the window layer or to reduce the recombination.

Amorphous alloys of silicon and carbon (a-Si$_x$C$_{1-x}$:H$_y$) are an promising alternative to standard a-Si$_x$:H$_y$. The introduction of carbon adds extra freedom to control the properties of the resulting material. An increasing concentration of carbon in the alloy *(x)* is used to widen the electronic gap between conduction and valence bands, in order to potentially increase the light efficiency of solar cells made with amorphous silicon carbide layers [60]. However, beside the ability to enhance the optical band gap and suppress absorption in the thin film emitter, it changes the electronic properties of hydrogenated amorphous silicon: the electronic semiconductor properties (mainly electron mobility), are badly affected by the increasing content of carbon in the alloy, due to the increased disorder in the atomic network. Bringing *(x)* to the opposite extreme of carbon concentration (100 %) results in amorphous carbon, or synthetic diamond-like films. Several studies are found in the scientific literature, mainly investigating the effects of deposition parameters on electronic quality (*cf.* [124–127]), but practical applications of amorphous silicon carbide in commercial devices are still lacking, in particular the use of a-Si$_x$C$_{1-x}$:H$_y$ in heterojunction solar cell devices. Therefore, one topic of this thesis is the investigation of plasma enhanced chemical vapor deposited layers in order to prove the feasibility to widen the optical band gap in emitters of heterojunction solar cells. In particular, this has been carried out by application of PECV deposited hydrogenated amorphous carbon-silicon alloys, a-Si$_x$C$_{1-x}$:H$_y$ and a-Si$_x$:H$_y$ layers.

4.2. Experiment: Hydrogenated amorphous carbon silicon alloys (a-SiC:H)

For the following investigations, ⟨100⟩ oriented, Czochralski (Cz), boron-doped (p-type), 1-4 Ωcm, 350 μm thick, one-side polished Si wafers are used. The resistivity employed is suitable for high performance of silicon solar cells. After a standard 2-step RCA process cleaning, followed by a 2 min diluted (2 %) hydrofluoric acid (HF) dip and deionized water rinsing, the wafer are inserted into a three chamber PECVD setup. The *n*-type a-Si$_x$C$_{1-x}$:H$_y$ and a-Si$_x$:H$_y$ layers are deposited on the polished side of the substrate using a direct PECVD (*cf.* section 3.1.1) reactor under conditions presented in Table 4.1. The total flux during deposition is kept constant at 36 sccm (sccm denotes standard cubic centimeter per minute at STP). The feedstock is set to 1 vol.% phosphine (PH$_3$) doping. The plasma exposures are performed at a frequency of 13.56 MHz and a power of 4.31 W with a substrate temperature of about 230 °C. The chamber pressure is adjusted to 300 mTorr. The deposition rates employed range from 0.5 to 2 Å·s^{-1}, depending on feed stock gas.

For preparation of the heterojunction solar cells, transparent conductive oxide (TCO) films (Indium-Tin-Oxide) are deposited on top of the a-Si$_x$C$_{1-x}$:H$_y$ layers, as described elsewhere (*cf.* section A.2). The sputtering time is set to 1100 s, so that a thickness of 80 nm is obtained. The front side metal grid (Ag, 4 μm) as well as the Al layer (4 μm) back-contact are fabricated by photo-

Table 4.1.: Precursor gas conditions for deposited n-type a-Si$_x$C$_{1-x}$:H$_y$ and a-Si$_x$:H$_y$ emitter layers - the total flux is kept constant at 36 sccm with a phosphine gas content of 1 vol.%; the silane flow is accordingly adjusted.

variation	sample no	CH$_4$ in feedstock (vol.%)	H$_2$ in feedstock (vol.%)
CH$_4$	SC1	0 - 40	0
CH$_4$	SC2	0 - 40	20
CH$_4$	SC3	0 - 45	30
H$_2$	SH1	0	10 - 40
H$_2$	SH2	15	0 - 50

lithography and e-gun evaporation. No back surface field (BSF) is applied and no intrinsic a-Si:H(i) layers are used for this investigations in order to keep the structure as simple as possible. Figure 4.1 sketches a cross-sectional view of the prepared heterojunction solar cell with wide band gap emitter, as prepared for the investigations in this chapter.

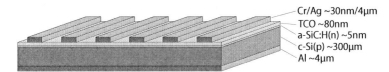

Cr/Ag ~30nm/4µm
TCO ~80nm
a-SiC:H(n) ~5nm
c-Si(p) ~300µm
Al ~4µm

Figure 4.1.: Cross-sectional view of the heterojunction solar cell.

4.3. Optical analysis

Spectroscopic ellipsometry (SE) measurements in the range between 1.0 and 4.5 eV are used to deduce the refractive index n and the extinction coefficient k as well as thickness of the deposited layers, following the analysis method described in section 3.2.3. To determine E$_G$ of the n-type a-Si$_x$C$_{1-x}$:H$_y$ and a-Si$_x$:H$_y$ films a dispersion model for amorphous material based on the absorption edge Tauc formula [63], and the quantum mechanical Lorentz oscillator model, proposed by Jellison and Modine (*cf*. [95]) is employed, as described in detail in section 3.2. The SE data - cos (Δ) and tan (Ψ) - are fitted assuming a three layer model - *surface roughness/a-Si film/substrate*. The op-

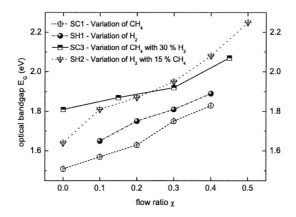

Figure 4.2.: Optical band gap (E_G) obtained from SE data for different gas compositions. χ represents the flow ratio for either methane to silane or hydrogen
to silane during decomposition. The phosphine flow is constant. The dotted lines are a guide to the eye.

tical gaps are then calculated from the ellipsometry data of films deposited
onto either transparent glass substrates or crystalline silicon wafer with and
without thermal silicon dioxide on top.

The widening of the optical band gap E_G of the deposited layers – differing
only in the precursor materials CH$_4$ and H$_2$ during decomposition – are shown
in Fig. 4.2 (details on obtaining the bandgap are described in section 3.2). The
PH$_3$ flow is kept constant at 1 vol.% of the total gas flux during deposition and
SiH$_4$ is accordingly adjusted to reach a total gas flux of 36 sccm.

The optical band gap E_{Tauc} is then determined in different states in each
composition range. The band gaps of the a-Si$_x$:H$_y$ films prepared by H$_2$ decomposition, as well as the a-Si$_x$C$_{1-x}$:H$_y$ films prepared by only CH$_4$ decomposition show a slight enhancement from 1.5 eV up to 1.82 and 1.88 eV, respectively. Likewise, the a-Si$_x$C$_{1-x}$:H$_y$ films - prepared by both hydrogen and
methane in the precursor materials - show a higher increase of the optical
band gap up to 2.08 eV for sample series SC3 and 2.27 eV for sample series
SH2.

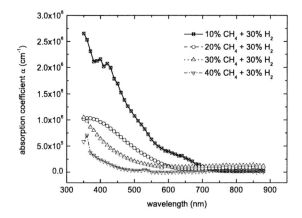

Figure 4.3.: The absorption coefficient, α, of the sample series SC3 calculated from SE data fit *vs.* the wavelength region from 300 to 880 nm.

For the a-Si_xC_{1-x}:H_y and a-Si_x:H_y thin films, which are responsible for the optical absorption, the deposition rate is determined to be 1-2 $\mathring{A} \cdot s^{-1}$. From the optical constants it is possible to derive the absorption coefficient α for each layer, independently from the desired layer thickness (*cf.* section 3.2). Fig. 4.3 shows the calculated optical absorption in the wavelength region from 300 to 880 nm for the sample series SC3 (30 % of hydrogen and varied amount of methane). The increasing carbon amount in the feed stock during deposition results in a distinct decrease of the absorption in the short wavelength region from 300 to 600 nm. Consistenly, the fraction of light transmitted to the wafer available for the solar conversion increases drastically.

For the sample series SH2, representing a 15 % of carbon and a varying hydrogen fraction in the feed stock, the wavelength dependent absorption coefficients are shown in Fig. 4.4. The same distinct decrease of the absorption as in Fig. 4.3 can be observed; here in the wavelength region up to 500 nm.

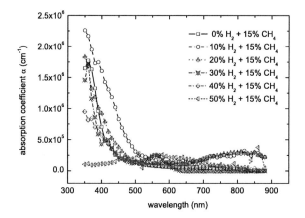

Figure 4.4.: The absorption coefficient, α, of the sample series SH2 calculated from SE data fit *vs.* the wavelength region from 300 to 880 nm.

4.4. Analysis of the incorporation of hydrogen and carbon by μ-Raman spectroscopy

To derive further information on the incorporation of the deposited layers, μ-Raman investigations have been carried out of films deposited on CZ c-Si wafers. The μ-Raman spectroscopy (μ-RS) setup is described elsewhere (*cf.* section 3.3). The excitation of the μ-RS is supplied by an Ar$^+$ ion laser using a wavelength of 488 nm and a power of 25 mW. The spot size of the focused laser light is ~1 μm. The spectra are collected at room temperature by a Peltier cooled charge coupled device detector. The resolution is limited to ~0.7 cm^{-1}. The spectra are recorded in the wavenumber range from 200 to 4600 cm^{-1}.

For μ-Raman measurements, samples with thicknesses of at least 100 nm are prepared on one-side polished wafers as well as on silicon oxide layers on wafers, whereas for heterojunction solar cells, layer thicknesses of ~5-10 nm are employed. It is assumed that the irregularity of the deposited layer thickness prepared for optical analysis and for the solar cell characteristics does not play a significant role for the layer composition of hydrogen and methane, which is investigated in the following.

The Raman spectra in the visible wavelength range corresponding to the wavenumber range of ~1650 to ~2650 cm^{-1} for various hydrogen and methane compositions are shown in Fig. 4.5. The spectra of samples with additional hydrogen incorporation seen in Fig. 4.5(a) are dominated by Si-H related bands at ~2100 cm^{-1}, indicating the presence of a-Si:H. The peak at 2320 cm^{-1} is due to the nitrogen in air. The band seen in Fig. 4.5(b) at ~2000 to ~2100 cm^{-1} broadens with increasing methane content in the precursor material. This can be attributed to the formation of dihydride (SiH$_2$) species with vibrational modes at ~2100 cm^{-1} in the deposited thin film, as reported in [128].

The broad band and the intensity of the spectra in Fig. 4.5(d) (30 % hydrogen with a increasing methane content) at ~2100 to ~2150 cm^{-1} enlarges with increasing methane content. This expansion is superimposed by the modes showing the typical signature of a-Si:H alloys [129] at ~2000 cm^{-1}. According to [128], at vibrational frequencies of ~2140 cm^{-1}, the structures observed in the Raman spectra of a-Si:H can be related to the increased presence of hydrogen, also shown in Fig. 4.5(c). The higher hydrogen content in the PECVD films can be explained by either trihydride SiH$_3$ bonds at ~2140 cm^{-1} [128] or it is due to the incorporation of C in the form of CH$_3$ groups [130], while Si is incorporated in the form of SiH or SiH$_2$ groups. The CH$_3$ radical is the dominant hydrocarbon radical in PECVD methane plasmas [131].

Further analysis shows that no C-C bonds are visible, as expected due to the bond stretching of sp^2 carbon atom pairs around ~1530 cm^{-1} (not shown in Fig. 4.5) [129]. The incorporation efficiency of C is lower than that of Si in a-Si$_x$C$_{1-x}$:H$_y$ alloys produced by PECVD from methane/silane mixtures [130] because methane possesses both a higher ionization energy and higher dissociation energy than silane.

The stretch modes of both Si-H bonds and C-H bonds of thin films deposited with both hydrogen and carbon fractions are analyzed: the Raman spectra of Si-H bonds in a-Si:H can be decomposed into two sub-peaks around ~2000 cm^{-1} and ~2100 cm^{-1} (shown in Fig. 4.5(d) as fit curves), which originate from isolated Si-H and larger (Si-H)$_x$ clusters located in large cavities or

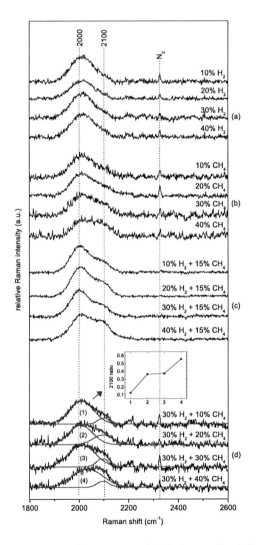

Figure 4.5.: Raman spectra after variation of (a) hydrogen (SH1), (b) methane (SC1) and (c, d) both hydrogen and methane (SH2, SC3) for 488 nm excitation in the wavenumber range from 1800 to 2600 cm^{-1}. Major peaks are labeled.

grain boundaries, respectively [132]. The ratio of the area of the sub-peak at ~2100 cm^{-1} to the area of whole Si-H peak, depending on the fraction of the H$_2$ or CH$_4$ in the precursor material during decomposition, is presented in the inset of Fig. 4.5. It is found that the ratio of hydrogen to silane in the precursor gas increases; the fraction of Si-H bonds in the form of (Si-H)$_x$ clusters tends to increase, too. This result is consistent with the optical band gap analysis in Fig. 4.2.

The Raman spectra exhibit more information about the carbon incorporation in the a-SiC:H alloys in the wavenumber range from 400 to 1000 cm^{-1}, presented in Fig. 4.6a. The first order Si peaks shown in Fig. 4.6a at ~490 cm^{-1} are due to the Si-Si peaks of the a-Si network in Si-rich alloys, while the peak at ~520 cm^{-1} is superimposed to the substrate c-Si signal. Apart from these peaks, the Raman spectra show two extra peaks at ~840 and ~890 cm^{-1}. These two peaks can be identified as peaks in the phonon density of states, which should be seen in a-SiC [133]. The peaks are extremly weak due to the small cross section of Si-C in the visible excitation [129]. Figure 4.6b reflects the Raman spectra in the high wavenumber range, with the C-H band peaks around ~2890 to ~2945 cm^{-1} [133], corresponding to sp^3 C-H stretching modes [129]. A detailed analysis of the ~2900 cm^{-1} band shows that with a methane content in the precursor material of about 30 % or less, the C-H bond peaks are extremely weak due to the incorporation of the C atoms surrounded by either Si atoms or C atoms. Samples with a methane content in the feed stock of about 40 % show C-H bonds at the ~2900 cm^{-1} band, indicating defect states.

The absence of a C-H signal for low methane content in the Raman spectra does not indicate the absence of C bondings in the film at all. In contrast, a distinct increase of the band gap due to carbon incorporation is observed, as shown in Fig. 4.2. In fact, if there would be no primary decomposition of methane, the incorporation of carbon in the film could only result from a chemical reaction between the 'hydrogen related defect species' (SiH, SiH$_2$, SiH$_3$, H) created by the plasma from the silane and the methane molecules. The absence of C-H signal for low methane content indicates a good passivation of carbon related defect states.

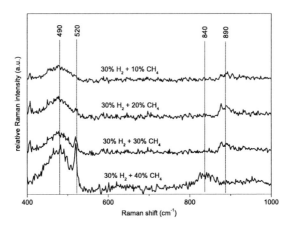

(a) Raman spectra in the wavenumber range from 400 to 1000 cm^{-1}.

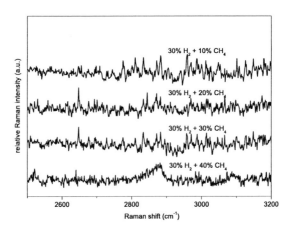

(b) Raman spectra in the wavenumber range from 2500 to 3200 cm^{-1}.

Figure 4.6.: Raman spectra of sample series SC3 (methane variation with 30 % hydrogen in the feedstock) for 488 nm excitation.

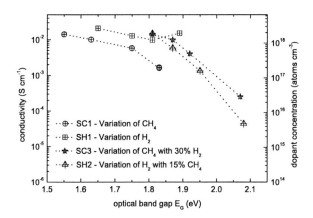

Figure 4.7.: Dark conductivity (left axis) and dopant concentration (right axis) *vs.* band gap (E_{Tauc}). The dotted lines are a guide to the eye.

4.5. Electrical characterization of the a-Si$_x$C$_{1-x}$:H$_y$ and a-Si$_x$:H$_y$ films

Fig. 4.7 exhibits the dark conductivity and dopant concentrations of the a-Si$_x$C$_{1-x}$:H$_y$ and a-Si$_x$:H$_y$ films. The dark conductivity is obtained by I-V measurements of the deposited layers on SiO$_2$ as insulator at room temperature, with a layer thickness of 100 nm. The dopant concentrations are calculated with an assumed electron mobility of $\mu = 0.5$ cm^2/V·s for all samples, *cf.* [123].

The addition of methane during the deposition of a-Si$_x$C$_{1-x}$:H$_y$ and a-Si$_x$:H$_y$ films reduces the optical losses in the light absorbing emitter of a heterojunction solar cell (*cf.* Fig. 4.3 and Fig. 4.4), but an increasing carbon content worsens the photoelectronic properties, such as photoconductivity, because the addition of carbon increases the density of electronic defect states. The tendency presented in Fig. 4.7 confirms the significant deterioration of dark conductivity for sample series SC3 resulting from C-H bondings at high methane concentrations. It can be seen that the addition of hydrogen (sample series SH1) does not affect the constant gradient of the conductivity, whereas the

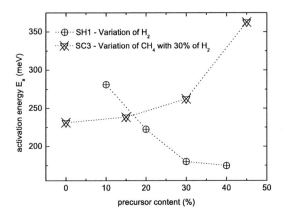

Figure 4.8.: Activation energy E_a *vs.* CH$_4$ or H$_2$ precursor content for sample series SC3 and SH1, respectively. The dotted lines are a guide to the eye.

addition of carbon in sample series SC3 decreases the conductivity, even with high hydrogen dilution. For further analysis, the activation energies E_a of the samples series SC3 and SH1 are obtained from Arrhenius plots taken at temperatures from 300 K to 425 K. The analysis depicts an increasing activation energy for sample series SC3 resulting from both increasing optical band gap and defect states through C atoms (see Fig. 4.8). In contrast, sample series SH1 shows an inverse behavior.

4.6. Effects on the solar cell properties

Some of the investigated a-Si$_x$C$_{1-x}$:H$_y$ films are deposited as emitters in heterojunction solar cells. The efficiency, η, of the heterojunction solar cells as well as short current density (J_{sc}) and open circuit voltage (V_{oc}) are derived from I-V measurements. Spectral response and reflection measurements led to the internal quantum efficiency (IQE).

To show the trade off between light absorption and defect density in the light absorbing emitter layer, IV-curves, measured under AM 1.5 conditions, are

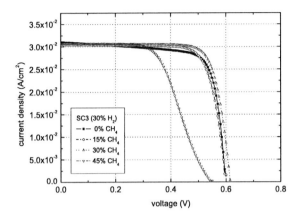

Figure 4.9.: Light I-V curves of sample series SC3.

presented in Fig. 4.9 and Fig. 4.10. They originate from a set of heterojunction
solar cells prepared using the process parameters of Table 1 for sample series
SC3.

The dark I-V curves reveal a suppression of the backward current density
by two orders of magnitude with increasing carbon concentration. The suppression of the backward current improves the V_{oc} (*cf*. Fig. 4.11) and hence
the solar cell performance until a carbon concentration of 30 % is reached.
Above 30 % carbon, the V_{oc} drops drastically. Figure 4.11 shows the results
for V_{oc} and J_{sc} taken from light and dark I-V curves for sample series SC3 compared to series SH2. It is found, that a certain amount of carbon and hydrogen
increases the V_{oc} as well as the J_{sc}.

To analyze the results presented in Fig. 4.11, the IQE from sample series
SH1 is presented in Fig. 4.12. For sake of simplicity, the fraction of methane
in the feed stock is taken as the fraction of the carbon. The effective diffusion
length, deduced from the near infrared wavelength range (880 nm - 970 nm)
is constant for all samples ($\sim$250 μm). From Fig. 4.12 can be deduced that the
spectral responses in the short wavelength range from 350 to 600 nm for the

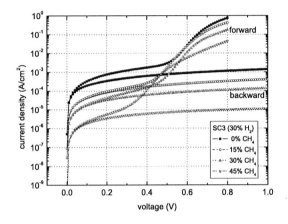

Figure 4.10.: Dark I-V curves of sample series SC3.

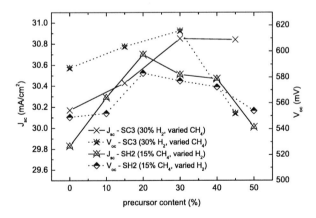

Figure 4.11.: Short circuit current J_{SC} (left axis) and open circuit voltage V_{OC} (right axis) vs. CH$_4$ or H$_2$ precursor content for sample series SC3 and SH2, respectively.

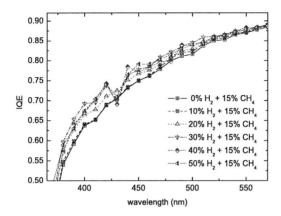

Figure 4.12.: IQE *vs.* wavelength in the wavelength region from 350 to 600 nm for sample series SH2 (variation of hydrogen fraction).

sample series SH2 tend to increase with increasing hydrogen fraction in the feedstock.

In Fig. 4.13, the obtained IQE for the sample series SC1 is presented. It can be concluded that the addition of C in the emitter layer decreases the spectral response slightly in the short wavelength range. This effect could be the consequence of the increased density of the electronic defect states due to the addition of C, though this deterioration is minimized by the hydrogen dilution, as seen in Fig. 4.12 from sample series SH2.

Figure 4.14 shows the measured efficiencies of the prepared heterojunction solar cells under AM 1.5 conditions for the sample series SC3 and SH2. In both cases, the optimum efficiency is obtained for films deposited with a hydrogen fraction of 30 % and a methane fraction of 20 %, which is a trade-off between low optical absorption and good electronic passivation. The relatively low efficiency with a maximum of 14.8 % can be explained due to the fact that in this work only the potential of the emitter and the decomposition of the different gas mixtures are investigated. Neither BSF, nor texturization nor passivation of the substrate are applied to achieve a high efficiency solar cell. A remark-

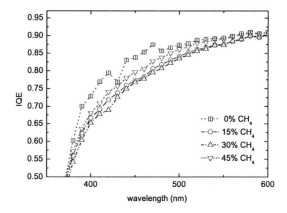

Figure 4.13.: IQE *vs.* wavelength in the short wavelength region from 350 to 600 nm for sample series SC1 (Variation of methane fraction).

ably low efficiency is observed for 45 % methane content in the feed stock with $\eta = 10.2$ %. This decrease is predictable as it is in line with the previously measured decrease of V_{oc}, resulting from the observed C-H bondings at high methane concentrations. None the less an improvement of the efficiency of 0.8 % for sample series SC3 due to hydrogenated carbon-silicon alloys is achieved.

4.7. Chapter summary

The influence of deposition parameters on the incorporation efficiency of both hydrogen and carbon in a-$\text{Si}_x\text{C}_{1-x}$:$\text{H}_y$ and a-Si_x:H_y films for use as emitters in heterojunction solar cells has been studied. It is found that the optical band gap E_G of those films can be tailored from 1.5 eV up to 2.3 eV with an appropriate addition of both methane as a carbon source and hydrogen during PECV deposition.

The μ-Raman spectra analyses have shown that additional methane as carbon source in the precursor material broadens the band at ~2000 to

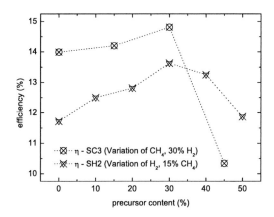

Figure 4.14.: Efficiencies of the sample series SH2 with increasing hydrogen content and fixed methane content of 15 % and of sample series SC3 with increasing carbon content and fixed hydrogen content of 30 % *vs.* H_2 or CH_4 precursor content.

~2100 cm^{-1}, which can be attributed to the formation of SiH_2 species. Compared to additional hydrogen without methane in the feedstock, only SiH related bonds are visible, indicating the typical a-Si spectra at ~2000 cm^{-1}. Addition of both, hydrogen and methane, shows an enhanced intensity and enlarged broad band from 2000 to 2150 cm^{-1}, indicating trihydride SiH_3 bonds and the incorporation of C in the form of CH_3 groups. Thus, Raman spectra at ~840 and ~890 cm^{-1} indicate carbon incorporation, which can be attributed to a-SiC. However, no C-C bonds are visible, which would lead to low conversion efficiency due to the sp^2/sp^3 bonding structure. Spectra in the high wavenumber range at ~2900 cm^{-1} exhibit peaks related to C-H bonds due to the high methane content in the precursor material.

Two sample series are investigated in more detail. First, 15 % methane with a defined hydrogen concentration in the precursor material and second, 30 % hydrogen with a defined methane concentration. For these sample series, the stretching modes of Si-H and C-H bonds are analyzed by decomposing the spectra into sub-peaks around ~2000 and ~2100 cm^{-1}.

Heterojunction solar cells, prepared with an appropriate addition of hydrogen and carbon, showed an increased V_{oc} as well as an increased J_{sc}. From IQE measurements (in the wavelength range from 350 nm to 550 nm) one can conclude that the addition of hydrogen leads to a slightly better internal quantum efficiency. The addition of carbon, however, decreases the IQE slightly in the short wavelength region. As a consequence, the conversion of photon energy into electric current implicates that it is not performed efficiently by just adding carbon, due to the sp^2/sp^3 bonding structure of C-C [60, 125, 126] as well as to the decreasing photoelectronic properties, such as photo-conductivity. Despite of that, the dark conductivity of the resulting thin films decreases with enhanced addition of carbon to the precursor material.

The trade-off between electrical defect density and optical absorption of the emitter layer results in a band gap of approximately 2.0 eV and a layer thickness of 5 to 10 nm. Light in the long-wavelength region passes this layer almost without loss; the optical losses in the blue spectral range are typically around 10 %.

The V_{oc} results of the cells discussed in this section appear to be quite poor, however, this is an expected behavior for any *p-n* or *n-p* heterojunction structure, as the interface state density caused by the doping materials, which attach to the c-Si surface during the deposition process, seem to deteriorate the junction properties significantly, *cf*. [9]. By the addition of carbon the optical band gap widens so that the suppression of light absorption in the window layer is reduced, but the density of defect states increases. Hydrogen dilution minimizes this deterioration. From our I-V measurements and the quantum efficiency measurements it is evident that the addition of carbon and hydrogen does contribute to the photo-generated current. It is recommended that by proper valency control of the optical band gap due to the hydrogenated amorphous carbon-silicon alloys a-Si$_x$C$_{1-x}$:H$_y$ J_{sc} and V_{oc} should be developed to a maximum efficiency of the solar cells. Also, with a non-doped a-Si:H layer sandwiched between the heterojunction, the hetero-interface is separated from the doped layer so that these defects caused by the doping materials might be avoided. This implies that the essence of the heterojunction solar cells consists in creating a good interface to avoid surface recombination [9], which will be discussed in chapter 6.

CHAPTER 5.

DEVELOPMENT OF WIDE-GAP HIGH CONDUCTIVE MICRO-CRYSTALLINE SILICON FILMS FOR USE AS EMITTER AND BSF IN HETEROJUNCTION SOLAR CELLS

5.1. Introduction

This chapter contains the development, optimization and analysis of wide-gap, high-conductive μc-Si:H layers (or a-Si:H layers which are grown close to the transition to micro-crystalline films) for use as emitter and back-surface-field (BSF) in heterojunction solar cell devices. It is well known that for i.e., thin film solar cells the PECVD conditions required to produce high quality a-Si:H are close to the region where micro-crystalline Si is deposited. With increasing crystallinity, the doping efficiency and carrier mobility increases. However, for the growth of those layers directly on c-Si, often an epitaxial growth at the interface is observed, reducing the effective lifetime and the V_{oc} of those devices [134].

In this work, for heterojunction solar cell device fabrication, the micro-crystalline layers are applied in combination with a underlaying amorphous surface passivating layer (cf. chapter 6), therefore an epitaxial growth of the μc-Si:H layers on c-Si is neglected.

The nature and corresponding properties of μc-Si:H films are already well developed (cf. [135, 136]). However, the PECV deposition of high quality micro-crystalline layers is not trivial, and depends strongly on the deposition parameters. Therefore, in this chapter the main attention is paid to the processes resulting in high-quality μc-Si:H films, i.e. highly conductive, and wide-gap films. In addition, the impact of post-annealing of the μc-Si:H(n) and μc-Si:H(p) layers is studied.

The p-type μc-Si:H (μc-Si:H(p)) layers are fabricated by decomposition of TMB (2 % diluted in H_2), SiH_4, and H_2, whereas the n-type μc-Si:H (μc-Si:H(n)) layers are deposited by decomposition of PH_3 (3 % diluted in SiH_4), SiH_4, and H_2. The deposition parameters, such as plasma frequency, deposition temperature, plasma power, deposition pressure are optimized in respect of the corresponding dark conductivity and resulting optical bandgap. The gas flow rates, in particular the influence of hydrogen dilution on the conductivity are investigated by Raman spectroscopy.

The terminology of 'micro-crystallinity' is not accurately defined in literature, as one can use the term 'micro-crystalline layer' or 'close to the transition to crystallinity' for a PECV deposited layer, which contains a fraction of micro-crystals. Hereafter, all μc-Si:H films or a-Si:H layers with a micro-crystalline fraction are referred to as 'micro-crystalline' films, whereas 'crystallinity' refers to the degree of structural order in silicon (usually specified as a volume percentage that is crystalline).

5.2. Experimental details

The μc-Si:H(p) and μc-Si:H(n) films are deposited by very high frequency (VHF) plasma enhanced chemical vapor deposition using gaseous precursors of SiH_4, H_2, and PH_3 or TMB for n^+ or p^+ doping, respectively. Depending on the analysis method, the films have been deposited on Corning glass 7059 or polished c-Si wafer substrates. The process parameters are as follows: plasma excitation frequency f_{plasma}, deposition pressure p_{dep}, gas concentration, heater temperature T_{heat} (deposition temperature T_{dep}), pre-heating time t_{heat}, plasma excitation power P_{rf}, and electrode inter-spacing d_e. The values of the varied process parameters are shown in Table 5.1.

The samples have been investigated using (i) Raman spectroscopy to analyze the resulting amorphous or μc-Si:H network of the deposited films (described in section 3.3), (ii) dark- and photo-conductivity measurements via the TLM method (described in section 3.4), and (iii) spectroscopic ellipsometry to investigate the corresponding optical properties such as transmission and optical bandgap (described in section 3.2).

Table 5.1.: PECVD conditions to fabricate n^+ μc-Si:H and p^+ μc-Si:H films. The dopant sources are PH$_3$ and TMB for n^+ and p^+ doping, respectively.

		PECVD conditions			
process	deposition temperature T_{dep} (°C)	precursor gas (sccm)	deposition pressure (mTorr)	plasma frequency (MHz)	plasma power (W/cm^2)
p^+	130–200	H$_2$: 0 − 200 SiH$_4$: 0 − 60 TMB : 0 − 40	200–1000	13.56–110	13.9–139.7
n^+	130–200	H$_2$: 0 − 200 SiH$_4$: 0 − 60 PH$_3$: 0 − 12	200–1000	13.56–110	13.9–139.7

For Raman spectroscopy, the crystalline volume fraction is a criterion to describe the silicon materials in its transition zone from amorphous to crystalline. Raman peaks at 480 cm^{-1} correspond to amorphous silicon, peaks at 520 cm^{-1} correspond to crystalline silicon structure. In Raman spectroscopy the standard peaks for crystalline and amorphous silicon appear at 520 cm^{-1} and 480 cm^{-1} wavenumbers, respectively. Therefore, the position of the Raman peaks for micro-crystalline silicon would determine prevalence of either crystalline or amorphous structure in the deposited films. Under this considerations, the peak at 500 cm^{-1} would indicate dominance of crystalline fraction in the examined samples, whereas the position of the peak close to 480 cm^{-1} attributes amorphous character of micro-crystalline film.

To determine the optical bandgap E_G of the micro-crystalline plasma deposited films, SE measurements have been carried out, as described in section 3.2. For analysis of the SE data, the dispersion model for amorphous material based on the absorption edge *Tauc* formula by *Davis and Mott* [63], and the quantum mechanical Lorentz oscillator model, proposed by *Jellison-Jr. and Modine* [95], have been employed. The SE data - cos (Δ) and tan (Ψ) - were fitted assuming a three to four layer model, depending on the substrate used (unless stated different): (from front to back) *surface roughness / amorphous/micro-crystalline film / corning-glass 7059*, or a four layer model *surface roughness / a-Si film / SiO$_2$ / c-Si* .

In case a c-Si substrate is used, the SiO$_2$ layer introduced serves as an op-

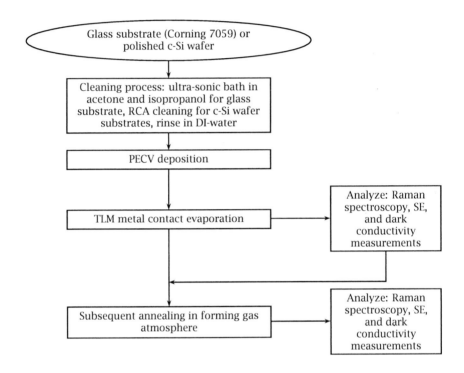

Figure 5.1.: Flowchart illustrating sample preparation and analysis of the prepared p$^+$ and n$^+$ μc-Si:H films.

tical separation layer. An accurate fit of very-thin films benefits from a sandwiched thermally grown silicon dioxide layer between crystalline wafer and amorphous/micro-crystalline plasma deposited layer to enhance the resolution.

Subsequently, the films have been annealed in forming gas atmosphere at temperatures in the range from 150 °C to 400 °C for 1 h. After annealing the impact on changes in the micromorph network and dark conductivity of the μc-Si:H films is investigated. Figure 5.1 summarizes the experiment in a flowchart.

5.3. p$^+$ doped μc-Si:H layer

5.3.1. Optimization of μc-Si:H(p) PECV deposition parameters

In anticipation of the dark conductivity results, the deposition pressure p_{dep} is optimal at 250 mTorr. For all following investigations presented below, the value of p_{dep} is kept constant. After preliminary tests the plasma excitation power P_{rf} and electrode inter-spacing d_e are set to 10 W and 19 mm, respectively. Those values ensure to obtain a homogeneous deposition. In the following experiments, a single parameter of plasma deposition condition is varied, whereas other parameters are kept constant.

5.3.1.1. Effect of hydrogen dilution

Figure 5.2 illustrates the dark conductivity as a function of H$_2$ dilution in sccm and as a partial pressure $\chi_H = [H_2]/([H_2]+[SiH_4])$. For thin μc-Si:H(p) films, reasonable dark conductivity values could be only achieved with high H$_2$ dilution. High dark conductivities up to 10 S/cm are obtained at H$_2$ dilution rates of more than 98 %. Below an χ_H of 98 %, the dark conductivity results in less than 10^{-4} S/cm. This result is further investigated by Raman spectroscopy.

In Fig. 5.3 the corresponding Raman spectra for μc-Si:H(p) films with various χ_H rates are shown. It is found that the fraction of microcrystallinity can be varied by H$_2$ dilution. At low χ_H rates amorphous films with peaks around 480 cm^{-1} are formed, whereas high H$_2$ dilution ($\chi_H > 98$ %) result in a microcrystalline character with peaks around 500 cm^{-1}. It can be concluded that the gain in dark conductivity with increasing H$_2$ content is directly correlated to the fraction of microcrystallinity.

5.3.1.2. Effect of deposition temperature

The deposition temperature can influence the fraction of micro-crystallinity. In Fig. 5.4 the Raman spectra for μc-Si:H(p) layers deposited at different temperatures in a range of $T_{heat} = 220$ - 400 °C. Increase of the deposition temperature results in a shift from amorphous (peaks around 480 cm^{-1}) towards

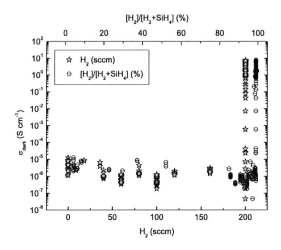

Figure 5.2.: Measured dark conductivity as a function of H_2 dilution in sccm (bottom) and as a partial pressure $[H_2]/([H_2]+[SiH_4])$ (top). High conductive layers are achieved with high H_2 dilution around 200 sccm.

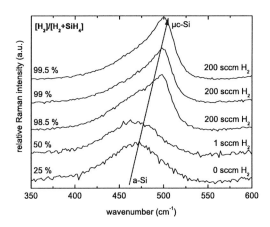

Figure 5.3.: Raman spectra of μc-Si:H(p) films for various H_2 dilutions. With low hydrogen dilution the films show an amorphous character (around 480 cm^{-1}), whereas very high hydrogen dilution exceeding 98 % results in a microcrystalline character (around 500 cm^{-1}). Taken from [137].

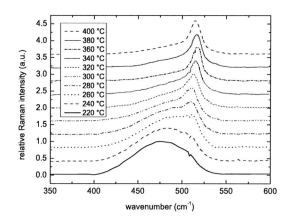

Figure 5.4.: Raman spectra of μc-Si:H(p) films deposited at various heater temperatures. Films deposited at low temperatures of 220 °C exhibit a wide band
at 480 cm^{-1}, typically for amorphous structure. Starting from 260 °C the
micro-crystalline fraction increases and reaches the absolute crystalline silicon value at the 520 cm^{-1} band. After [88].

micro-crystalline (peaks around 500 cm^{-1}) structure of the films. Films deposited at low temperatures of 220 °C exhibit a wide band at 480 cm^{-1}, typically for amorphous structure. Starting from 260 °C the micro-crystalline
fraction increases and reaches the absolute crystalline silicon value at the
520 cm^{-1} band.

Figure 5.5 displays the dark conductivity values as a function of heater temperature during deposition. An optimum temperature appropriate for high
conductivity is found as T_{heat} = 280 °C for the PECVD setup used. Of course
this temperature depends on the electrode distance. After Fig. 3.4 in section 3.1.1, the value of T_{heat} = 280 °C corresponds to a deposition temperature
$T_{dep} < 140$ °C. For this deposition temperature conditions, dark conductivity
values as high as 10 S/cm are obtained.

Figure 5.6 shows the optical bandgap obtained by SE measurements as a
function of heater temperature T_{heat}. It can be concluded that with decreasing deposition temperature the optical bandgap increases up to 2.04 eV at

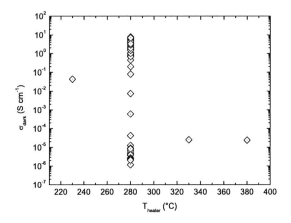

Figure 5.5.: Measured dark conductivity as a function of the heater temperature (T_{heat}). The optimum temperature appropriate for high conductivity turned out to be 280 °C for the PECVD setup used. The optimum temperature is a function of the electrode distance.

T_{heat} = 220 °C. In comparison with the containing amorphous fraction in the μc-Si:H(p) films, deduced from Fig. 5.4, one can conclude that a increasing amorphous fraction results in a increase of the optical bandgap.

5.3.1.3. Influence of pre-heating time

For investigations of the pre-heating time (t_{heat}) influence, samples are deposited with identical PECVD process parameters (T_{heat} = 280 °C), varying only in the pre-heating time in the chamber prior deposition. The influence of the pre-heating time on the resulting conductivity of the μc-Si:H(p) films is depicted in Fig. 5.7. In section 3.1.1 is already mentioned that the deposition temperature saturates after a pre-heating time of 60 min. Therefore, the influence of the pre-heating time changes only slightly the resulting dark conductivity values, as depicted in Fig. 5.7. Samples pre-heated for 15 min (corresponding to a T_{dep} = 140 °C) exhibit dark conductivity values of around 4 S/cm; with increasing t_{heat} the dark conductivity decreases to 3 S/cm. The

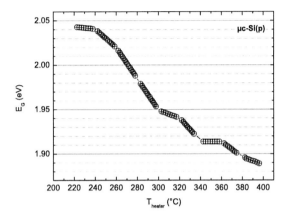

Figure 5.6.: Optical bandgap obtained from SE measurements for μc-Si:H(p) films as a function of heater temperature. With decreasing deposition temperature the optical bandgap increases up to 2.04 eV at T_{heat} = 220 °C. After [88].

sample temperature T_{dep} after 60 min pre-heating reaches its maximum at 160 °C. Thus, the decrease of dark conductivity at longer pre-heating times can be understood as temperature assisted distortion of the film material.

5.3.1.4. Effect of TMB concentration

Dark conductivity measurements also provides information on the TMB concentration effect. Figure 5.8 exhibits the change in the dark conductivity of μc-Si:H(p) films as a function of TMB concentration (χ_{TMB} = [TMB]/([TMB]+[SiH$_4$])). The dark conductivity increases with increasing TMB concentration till a maximum at χ_{TMB} = 2.62 %, corresponding to approximatly 30000 ppm. Note that the gaseous TMB used is diluted 2 % in hydrogen. With further increase of TMB above 2.75 %, the dark conductivity decreases due to interstitial boron, acting as recombination centers.

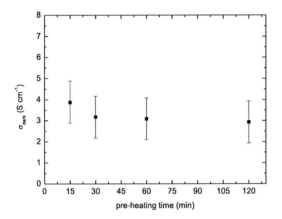

Figure 5.7.: Measured dark conductivity of μc-Si:H(p) films deposited on a glass substrate as a function of pre-heating time t_{heat} in the PECVD chamber. A slight influence of t_{heat} on the resulting dark conductivity can be noticed.

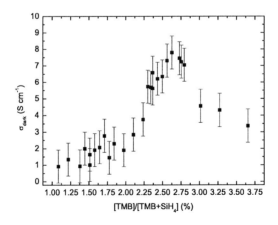

Figure 5.8.: Dark conductivity of p^+ μc-Si:H films as a function of TMB concentration in the feedstock during PECV deposition. A maximum is obtained at 2.62 % χ_{TMB}. The dilution of TMB of 2 % in H_2 is substracted.

5.3.2. Impact of post-annealing of μc-Si:H(p) films on electrical properties

In this experiment the impact of post-annealing of the deposited μc-Si:H(p) films is analyzed by dark conductivity measurements as a function of TMB concentration and annealing temperature T_{ann}. As shown above, the TMB concentration of 2.62 % results in the highest measured dark conductivity of 8 S/cm for μc-Si:H(p) layers. For the simultaneously investigations of dark conductivity dependence of annealing temperature and TMB concentration, the change in conductivity as a function of annealing temperature and TMB concentration is displayed in Fig. 5.9a. It can be seen that at annealing temperatures above 200 °C the dark conductivity increases drastically. It is known, that post-annealing of the μc-Si:H films or a-Si:H films with a micro-crystalline fraction results in an activation of the boron atoms, which formerly was bound in form of B-H-Si. The annealing step might release the hydrogen from this B-H-Si bound, leaving activated B-Si bounds, *cf. Rath et al.* [138].

At annealing temperatures above T_{ann} = 375 °C, the conductivity of the μc-Si:H(p) layers decreases. This effect can be attributed to hydrogen effusion resulting in a higher defect level. For better readability of the 3-D plot displayed in Fig. 5.9a, Fig. 5.9b displays the dark conductivity as a function of annealing temperature in a two-dimensional scale. The conductivity increases above T_{ann} > 200 °C drastically with increasing annealing temperature. This results suggests that boron atoms neutralized in the as-deposited μc-Si:H(p) films are activated as T_{ann} increases.

5.4. n$^+$ doped μc-Si:H layer

5.4.1. Optimization of μc-Si:H(n) PECV deposition parameters

In anticipation of the dark conductivity results, the deposition pressure p_{dep} is optimal at 350 mTorr. For all following investigations, the value of p_{dep} is kept constant. After preliminary tests the plasma excitation power P_{rf} and electrode inter-spacing d_e are set to 10 W and 19 mm, respectively. Those values ensure to obtain a homogeneous deposition. In the following experi-

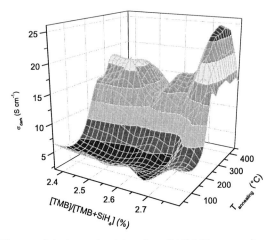

(a) Measured dark conductivity of μc-Si:H(p) films as a function of annealing temperature and TMB concentration in the feedstock during PECV deposition.

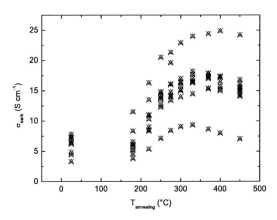

(b) For better readability a two-dimensional view of the above diagram.

Figure 5.9.: Measured dark conductivity of μc-Si:H(p) films deposited on glass substrates as a function of annealing temperature. The samples have been post-annealed for 1 h at various temperatures.

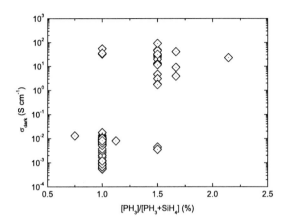

Figure 5.10.: Dark conductivity of μc-Si:H(p) films as a function of PH_3 concentration in the feedstock during PECV deposition. A maximum is obtained at 1.5 % χ_{PH_3}. The dilution of 3 % ph in SiH_4 is substracted.

ments, a single parameter of plasma deposition condition is varied, whereas all other parameters are kept constant.

5.4.1.1. Effect of phosphine concentration

Since the phosphine concentration mainly influences the doping efficiency and resulting carrier mobility, the effect of PH_3 doping concentration in the feedstock during composition in the PECVD chamber is investigated in the following. For preparation of the samples, f_{plasma} and T_{heat} are set to 110 MHz and 320 °C, respectively. Figure 5.10 shows the change in the dark conductivity of μc-Si:H(n) films as a function of PH_3 concentration (χ_{PH_3} = $[PH_3]/([PH_3]+[SiH_4])$).

The influence of PH_3 doping appears to be less pronounced than the influence of TMB concentration as shown in Fig. 5.8 (*cf*. section 5.3.1.4). Highest dark conductivity values are obtained with PH_3 dilution around χ_{PH_3} = 1.5 %. Note that PH_3 is diluted 3 % in SiH_4, and the corresponding fraction of SiH_4 is added to the total SiH_4 gas flux.

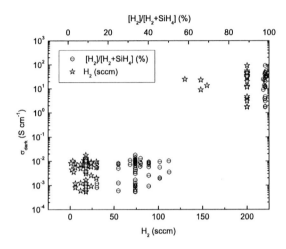

Figure 5.11.: Measured dark conductivity as a function of H_2 dilution in sccm (bottom) and as a partial pressure $[H_2]/([H_2]+[SiH_4])$ (top). High conductive layers are achieved solely with high H_2 dilution around 200 sccm.

5.4.1.2. Effect of hydrogen dilution

The hydrogen dilution has a strong influence on the crystallinity and is therefore responsible for the corresponding dark conductivity. The effect of H_2 dilution in the μc-Si:H(n) films is studied in the following. Figure 5.11 illustrates the dark conductivity as a function of H_2 dilution in sccm and as a partial pressure $\chi_H = [H_2]/([H_2]+[SiH_4])$. The f_{plasma} is set to 110 MHz, $T_{heat} = 320$ °C. It can be concluded that high dark conductivities up to 100 S/cm are obtained at high H_2 dilution rates of more than $\chi_H > 98$ %. It is obvious that a high H_2 dilution is indispensable for high conductivity layers. H_2 dilution of less than 150 sccm results in $\sigma_{dark} < 10^{-2}$ S/cm. This result is further investigated by Raman measurements.

In Fig. 5.12 the corresponding Raman spectra for μc-Si:H(p) films with various χ_H rates are displayed. It is found that the fraction of micro-crystallinity can be varied by H_2 dilution: low χ_H rates result in a rather amorphous to micro-crystalline character (close to the transition of micro-crystalline) with peaks around 480 - 500 cm^{-1}, whereas high H_2 dilution ($\chi_H > 98$ %) result in a

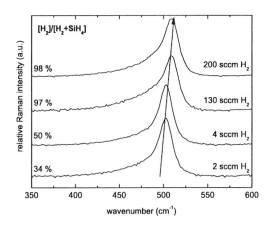

Figure 5.12.: Raman spectra of μc-Si:H(n) films for various H_2 dilutions. With low H_2 dilution the films show a less pronounced micro-crystalline character, whereas very high H_2 dilution exceeding 98 % results in a clear micro-crystalline to crystallinity character. Taken from [137].

micro-crystalline character with peaks around 500 - 520 cm^{-1}. These results are in line with the results achieved for μc-Si:H(p) layers: The fraction of H_2 and therefore the fraction of microcrystallinity directly corresponds to the resulting dark conductivity.

5.4.1.3. Effect of plasma frequency

The applied plasma frequency is an important factor to obtain high doping efficiency of the μc-Si:H(n) films. The films are deposited using T_{heat} = 320 °C. Optimal gas concentrations, such as a H_2 dilution of 200 sccm, and a PH$_3$ concentration of χ_{PH_3} = 1.5 % are used for all samples. Figure 5.13 displays measured dark conductivities of μc-Si:H(n) films deposited at plasma frequencies of 13.56 MHz, 70 MHz, and 110 MHz. Dark conductivities of σ_{dark} = 100 S/cm are obtained with a plasma frequency of 110 MHz. Lower plasma frequencies applied do not lead to such high σ_{dark} values. At plasma frequencies of 13.56 MHz the value of the dark conductivity σ_{dark} appears by more than four

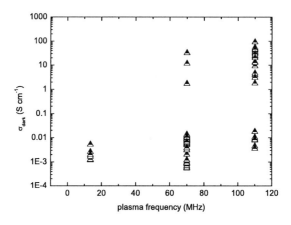

Figure 5.13.: Measured dark conductivity as a function of applied plasma frequency. It is obvious that the doping efficiency and carrier mobility increases with higher plasma frequencies.

orders of magnitude lower compared to one taken at 110 MHz. At f_{plasma} = 70 MHz the values of the dark conductivity are higher, but still by factor of 2 lower than the one obtained at 110 MHz.

Raman spectroscopy exhibit the correlation of doping efficiency or carrier mobility to applied plasma frequency. Figure 5.14 shows the Raman spectra for μc-Si:H(n) films fabricated with two different plasma frequencies (namely those plasma frequencies resulting in high σ_{dark} values). Films deposited at 70 MHz show a micro-crystalline character with a high amorphous fraction (broad peaks at around 480 cm^{-1}). Whereas films deposited at 110 MHz are of fully micro-crystalline character, as no peaks are observed at wavenumber lower than 500 cm^{-1}. It is obvious that the doping efficiency (and resulting carrier mobility) increases with higher plasma frequencies, and that a micro-crystalline network is needed for σ_{dark} values as high as 100 S/cm. Based on that one can conclude that a plasma frequency of 110 MHz is the optimal frequency for μc-Si:H film deposition.

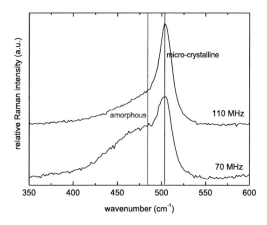

Figure 5.14.: Raman spectra for μc-Si:H(n) films fabricated with two different plasma frequencies. Films deposited at 70 MHz show a micro-crystalline character with a high amorphous fraction, whereas films deposited at 110 MHz are of fully micro-crystalline character. Taken from [137].

5.4.1.4. Effect of deposition temperature

The effect of deposition temperature is investigated by measuring the dark conductivity as a function of heater temperature T_{heat}. The heater temperature directly influences the deposition temperature T_{dep} (*cf.* section 3.1.1, Fig. 3.4). For this experiment, f_{plasma} is set to the optimum value of 110 MHz. Constant gas concentrations, such as a H_2 dilution of 200 sccm, and a PH_3 concentration of $\chi_{PH_3} = 1.5$ %, are used for all samples. Figure 5.15 illustrates the dependence of measured dark conductivity for μc-Si:H(n) layers as a function of heater temperature in a range from 250 °C to 400 °C. An optimum heater temperature of $T_{heat} = 320$ °C is found at highest $\sigma_{dark} = 100$ S/cm. This heater temperature is corresponding to a deposition temperature of $T_{dep} = 155$ °C. At lower or higher heater temperatures, significantly lower dark conductivities of $\sigma_{dark} < 10^{-1}$ S/cm are detected. Obviously, the doping efficiency does strongly depend on the deposition temprature, which is optimal at $T_{dep} = 155$ °C.

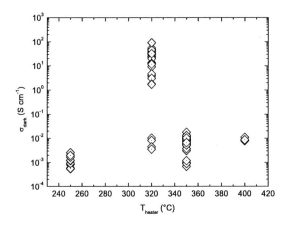

Figure 5.15.: Measured dark conductivity as a function of heater temperature during deposition.

Figure 5.16 displays the optical bandgap obtained by SE measurements as a function of T_{heat}. For a $T_{heat} = 280$ °C, a $E_G = 1.988$ eV has been determined. With increasing $T_{heat} = 320$ °C the bandgap decreases slightly to $E_G = 1.98$ eV. This result is consistent with results achieved for PECV deposited amorphous silicon, where a increasing deposition temperature results in lower hydrogen fraction in the deposited layers, leading to a lower bandgap, *cf*. [139].

5.4.2. Impact of post-annealing of μc-Si:H(n) films on electrical properties

Post-annealing of μc-Si:H(n) films has been carried out for various annealing temperatures, each annealing process equals 1 h in forming gas atmosphere. The samples are post-annealed after deposition, the values of σ_{dark} are then measured, and the sample is annealed again. The influence of the temperature is assumed to be significant higher compared to the time-factor. The data of post-annealed μc-Si:H(p) films confirm an activation of the dopant, boron, *cf*. Fig. 5.9. This activation of the dopant boron resulted in a significant increase

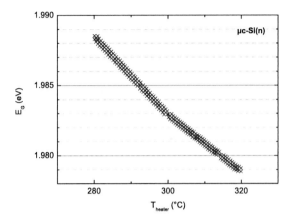

Figure 5.16.: Optical bandgap obtained from SE measurements for μc-Si:H(n) films as a function of heater temperature. After [88].

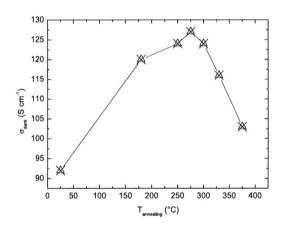

Figure 5.17.: Measured dark conductivity as a function of annealing temperature of μc-Si:H(n) films.

Table 5.2.: PECVD conditions for optimized n^+ μc-Si:H and p^+ μc-Si:H films. The dopant sources are PH$_3$ (3 % in SiH$_4$) and TMB (2 % in H$_2$) for n^+ and p^+ doping, respectively.

	PECVD conditions					
process	T_{heat} (T_{dep}) (°C)	precursor gas (%)	process pressure (mTorr)	plasma frequency (MHz)	plasma power (W/cm^2)	σ_{dark} (S/cm)
p^+	280 (140)	r_h = 99.01 r_{tmb} = 2.62	250	110	69.9	25
n^+	320 (155)	r_h = 98.07 r_{ph} = 1.5	350	110	69.9	130

of σ_{dark}, in case of the μc-Si:H(p) films. However, no significant increase could be detected for the optimized μc-Si:H(n) films. Figure 5.17 shows the impact of post-annealing on the resulting σ_{dark} as a function of annealing temperature. In contrary to the behavior of the annealed μc-Si:H(p) samples, the conductivity of the μc-Si:H(n) films exhibits a slight increase from 100 S/cm to 128 S/cm at T_{ann} in the range of 275 °C. With further increase of the annealing temperature up to 330 °C, the value of the dark conductivity drops to 102 S/cm. Although the value of the dark conductivity degrade at T_{ann} > 330 °C, it is still superior to the values of not annealed μc-Si:H(n) films. In conclusion, the μc-Si:H(n) films benefit from post-annealing in a temperature range of 180 °C < T_{ann} < 300 °C.

5.5. Chapter summary

According to the film properties discussed in this chapter, the p^+ and n^+ μc-Si:H films are likely to be suitable for use as emitter and BSF in a heterojunction solar cell device. They indicted high transparency to suppress absorption, and high conductivity when annealed at the optimum temperature.

Table 5.2 displays the PECVD parameters for μc-Si:H(p) and μc-Si:H(n) layers, optimized in terms of dark conductivity and optical transmission.

For p^+ μc-Si:H layers, the dark conductivity increases with increasing TMB concentration till a maximum at χ_{TMB} = 2.62 %, corresponding to approximatly 30000 ppm. An optimum heater temperature appropriate for high conductivity is found at T_{heat} = 280 °C for the PECVD setup used, corresponding to a deposition temperature of T_{dep} = 140 °C. It is found that the fraction of microcrystallinity can be varied as a function of hydrogen dilution. Low χ_H rates result in a larger fraction of amorphous structures inside the deposited film with Raman peaks around 480 cm^{-1}, whereas high H$_2$ dilution (χ_H > 98 %) result in a more micro-crystalline character with peaks around 500 cm^{-1}. The μc-Si:H(p) layers indictaed at optimal PECV deposition parameters high conductivities of σ_{dark} = 10 S/cm (corresponding to 0.1 Ωcm).

The n$^+$ μc-Si:H films indicated significant higher dark conductivity values (with optimum process parameters approximately σ_{dark} = 100 S/cm) compared to the σ_{dark} values of p$^+$ μc-Si:H films. The doping concentration for the highest σ_{dark} is found to be approximately χ_{PH_3} = 1.5 %. It can be concluded that the fraction of micro-crystallinity can be varied the plasma frequency, and is also influenced by the deposition temperature and hydrogen dilution. It is found that the doping efficiency and carrier mobility are optimal at VHF plasma frequency of 110 MHz, and with high hydrogen dilution of χ_H = 98 %.

The dark conductivity σ_{dark} and hence the carrier mobility are improved by means of thermal annealing for both n$^+$ and p$^+$ μc-Si:H films. The μc-Si:H(p) films annealed with temperatures in the range of 200 °C < T_{ann} < 375 °C are appropriate to obtain the high dark conductivity values (with optimum process parameters results in approximately σ_{dark} = 25 S/cm). By means of postannealing of the μc-Si:H films or a-Si:H films with a micro-crystalline fraction, an activation of the boron atoms occurs, which was formerly bound in form of B-H-Si. The annealing step might release the hydrogen from this B-H-Si bound, leaving activated B-Si bounds. For μc-Si:H(n) films an annealing temperature in the range between 180 °C < T_{ann} < 300 °C is found to be an optimum temperature to improve dark conductivity up to σ_{dark} = 130 S/cm (corresponding to 0.0077 Ωcm).

CHAPTER 6.

SURFACE RECOMBINATION OF CRYSTALLINE SILICON PASSIVATED WITH PECVD AMORPHOUS SILICON SUB-OXIDE

6.1. Introduction

In the research field of crystalline silicon (c-Si) solar cells, electronic surface passivation has been recognized as a crucial step to achieve high conversion efficiencies. High bulk and surface recombination rates are known to limit the open circuit voltage and to reduce the fill-factor of photovoltaic devices [26, 33]. The suppression of surface recombination by applying a surface passivation scheme is thereby one of the basic prerequisites to obtain high efficiency solar cells. This becomes particularly true for heterojunction solar cells: featuring an abrupt discontinuity of the crystal network at the crystalline silicon (c-Si) surface to the amorphous emitter (a-Si:H) usually results in a large density of defects in the bandgap due to a high density of dangling bonds [60]. These defects at the hetero-interface often induce detrimental effects on the solar cell performance, cf. [140]. Although the electrical field can reduce the recombination near the hetero-interface, the junction properties are still governed by the interface state density. Therefore, in order to obtain high efficiency solar cells it is essential to reduce the interface state density [9].

Passivation schemes commonly used in photovoltaic applications utilize silicon dioxide (SiO_2) [26], and silicon nitride (SiN_x) [27, 28], but also intrinsic amorphous silicon (a-Si:H(i)) [6], and amorphous silicon carbide (a-SiC:H) [29]. Thermally grown silicon oxide has shown excellent surface passivation properties (cf. [26]), resulting in a very low state density. However, the growth implies a high temperature application (~1050 °C), and it suffers from long term UV-instability. Low-temperature processing sequences are based mainly on passivation with silicon-nitride (SiN$_x$) [27], amorphous silicon films [6], or amorphous silicon carbide [29] or stacks of those. The SiN$_x$ films are silicon rich and this fact brings along several drawbacks: the passivation quality depends strongly on the Si doping type and level used; the films show a considerable absorption in the ultra-violet range of the solar spectrum, leading to a reduction of the J$_{sc}$; the etch rate of those films is extremely low, hindering the local opening of the SiN$_x$, which make them non applicable for heterojunction solar cells. Especially for heterojunction solar cell applications, a-Si:H(i) has attracted the photovoltaic community due to the success of HIT™(Heterojunction with Intrinsic Thin-Layer) cells [8, 9, 82]. The a-Si:H(i) films can be grown by PECV deposition at low temperatures (< 200 °C). However, *Fujiwara and Kondo* [18] stated that the growth of a-Si:H at temperatures T$_{dep}$ > 130 °C often leads to an epitaxial layer formation on the c-Si, reducing the solar cell performance. Also, due to the inherent strong blue light absorption, only ultra-thin a-Si(i):H films can be allowed to prevent losses. Focusing on the junction fabrication techniques of a-Si/c-Si solar cells using a low-temperature PECVD technique, passivation of the surface regions of the cell avoiding high-temperature cycling becomes an important issue. High-transparent PECV deposited hydrogenated amorphous silicon suboxides (a-SiO$_x$:H), grown at low temperatures, represent a material system suitable for this application and are quite an attractive alternative to standard a-Si:H incorporated in a-Si/c-Si heterojunction solar cells. In this chapter, the applicability of PECV deposited a-SiO$_x$:H films as passivation layers for any c-Si based solar cell device is investigated. For comparison, intrinsic amorphous silicon films (a-Si:H(i)) with no oxygen content are identically processed.

6.2. Previous work

6.2.1. SiO$_2$ and SiN$_x$ passivation schemes

As described in section 6.1, standard passivation schemes include high quality SiO$_2$ and SiN$_x$ films for high efficiency device fabrication. Up to now, record values of effective lifetimes are established for SiO$_2$ and SiN$_x$

Table 6.1.: Exemplarily values for measured effective lifetimes and the corresponding upper bound (S_{eff} by assuming ($\tau_{bulk} \to \infty$)) on the surface recombination velocity (SRV) at Si/SiO_2 and Si/SiN_x interfaces. Values published
by *Kerr and Cuevas* in [26, 27].

dopant type	resis- tivity (Ωcm)	wafer width (μm)	passi- vation method	τ_{eff} @ maximum measured (ms)	upper bound $S_{eff-max}$ (cm/s)	τ_{eff} @ $1 \cdot 10^{15}$ cm^{-3} (ms)	upper bound S_{eff} (cm/s)
n	90	265	SiO_2	29	0.46	16.2	0.82
n	1.5	285	SiO_2	6	2.4	5.2	2.74
n	0.6	590	SiO_2	1.1	26.8	1.1	26.8
p	150	400	SiO_2	32	0.63	22.5	0.88
p	1	400	SiO_2	1.7	11.8	1.7	11.8
n	90	275	SiN_x	10	1.38	3.79	3.63
n	1.5	305	SiN_x	2.15	7.2	2	7.63
n	0.6	350	SiN_x	0.96	18.3	0.96	18.3
p	150	440	SiN_x	9.60	2.3	6.2	3.54
p	1	400	SiN_x	1.05	21.8	1.05	21.8

by *Kerr and Cuevas* published in [26, 27]: the highest effective lifetimes
(τ_{eff}) which incorporates all the bulk and surface recombination processes,
previously reported for crystalline silicon appears to be τ_{eff} = 29 ms and
τ_{eff} = 32 ms, corresponding to the lowest S_{eff} with 0.46 cm/s and 0.625 cm/s
for a 90 Ωcm FZ n-type material and a 150 Ωcm FZ p-type material, respectively, passivated with annealed SiO_2. The measured effective lifetimes for
oxide-passivated (SiO_2), annealed n-type and p-type wafers, as well as n-type
p-type silicon passivated with stoichiometric SiN_x films are shown in Fig. 6.1.
Note, that the values given are obtained using the effective lifetime at its maximum value, neglecting the choice of a discretionary injection level. However,
the SRV for higher doped material, i.e., 1.5 Ωcm and 0.6 Ωcm n-type material,
increased to 2.4 cm/s and 26.8 cm/s, respectively, for wafers passivated with
SiO_2, as shown in Fig. 6.1a. Table 6.1 displays a comparison of the record
passivation values of τ_{eff} and S_{eff} achieved by *Kerr and Cuevas*.

From the results in Table 6.1 it is obvious that the values for the measured
τ_{eff} and the corresponding S_{eff} strongly depend on the chosen value for the
mean carrier density (MCD), and the c-Si wafer doping and resistivity used for
the investigation of passivation quality.

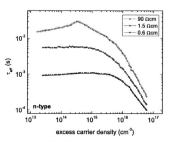

(a) Measured effective lifetimes for oxide-passivated (SiO$_2$), annealed n-type wafers.

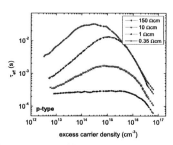

(b) Measured effective lifetimes for oxide-passivated (SiO$_2$), annealed p-type wafers.

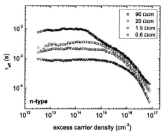

(c) Measured effective lifetimes for n-type silicon passivated with stoichiometric SiN$_x$ films.

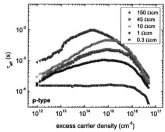

(d) Measured effective lifetimes for p-type silicon passivated with stoichiometric SiN$_x$ films.

Figure 6.1.: Previously published record effective lifetime values for SiO$_2$ and SiN$_x$ passivation schemes. Taken from [26, 27].

6.2.2. a-Si:H(i) passivation scheme

Excellent surface passivation abilities of non-doped a-Si:H(i) films on c-Si has been intensively investigated by other research institutes and confirmed by various methods, cf. [15, 141, 142]. *Dauwe et al.* [143] published very low surface recombination velocities on *p*- and *n*-type silicon wafers passivated with hydrogenated amorphous silicon films. Silicon material of both conduction types can be effectively passivated. *Wang et al.* [144] investigated a-Si:H/c-Si hetero-interfaces, stating that they necessitate an immediate a-Si:H deposition and an abrupt and flat interface to the c-Si substrate. High temperature growth of a-Si:H(i), however, often leads to epitaxial layer formation on c-Si, which in turn reduces the solar cell performance severely. Thus, this epitaxial growth narrows the process window for the a-Si:H(i) layer growth and makes the solar cell optimization more difficult [18]. Physical and structural properties of a-Si:H, however, strongly vary with the growth conditions, and the characteristics of the a-Si:H/c-Si hetero-interface still remain ambiguous.

6.2.3. a-SiC:H passivation scheme

Passivation schemes using a-SiC:H are intensively investigated by i.e., *Vetter et al.* [145] and *Martin et al.* [29]. An a-SiC:H passivation of 3.3 Ωcm *p*- and 1.5 Ωcm *n*-type c-Si results in $\tau_{eff} > 1$ ms. Using the same model as *Garin et al.* [146] leads to the identification of a field-effect passivation mechanism: fixed positive charge for *p*-type c-Si, fixed negative charge for *n*-type c-Si. *Vetter et al.* [147] investigated the effect of amorphous silicon carbide layer thickness on the passivation quality of crystalline silicon surfaces.

6.2.4. a-SiO$_x$:H passivation scheme

A novel passivation scheme using amorphous silicon sub-oxides(a-SiO$_x$:H) is introduced during this work (cf. also *Mueller et al.* [148]). The basic properties of PECV deposited a-SiO$_x$:H are discussed already by *Das et al.* [149] and *Janotta et al.* [150]. Recently, *Yamamoto et al.* [151] and *Sritharathikhun et al.* [152] published the use of a-SiO$_x$:H as an passivation scheme, although with unsatisfactory results.

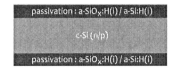

Figure 6.2.: Sketch of the sample structure for lifetime measurements.

6.3. Experimental details for a-SiO$_x$:H

Intrinsic amorphous hydrogenated silicon (a-Si:H(i)) is generally grown by decomposition of H$_2$ and SiH$_4$, whereas amorphous hydrogenated silicon sub oxides (a-SiO$_x$:H(i)) need an additional oxygen source. The decomposition of carbon dioxide (CO$_2$) as oxygen source, hydrogen (H$_2$), and silane (SiH$_4$) as precursor gases leads to the designated a-SiO$_x$:H films. For this purpose a three chamber high-frequency plasma enhanced chemical vapor deposition setup, as described in section 3.1.1, is used.

The oxygen content and the resulting optical band gap E$_{04}$ of the a-SiO$_x$:H films are controlled by varying the CO$_2$ partial pressure [CO$_2$]/([CO$_2$]+[SiH$_4$]). Therefore, all oxygen concentrations quoted in this work - unless stated differently - refer to the ratio χ_{ox} = [CO$_2$]/([CO$_2$]+[SiH$_4$]).

To determine E$_G$ and the optical constants, the films are deposited onto c-Si wafers with thermal SiO$_2$ (1 000 Å). Corning 7059 glass is used for deposition for μ-Raman spectroscopy. For lifetime measurements, the a-SiO$_x$:H are directly deposited onto double-side polished FZ wafers of different doping type and levels.

The typical sample configuration to determine the passivation quality of the a-SiO$_x$:H films is a symmetric a-SiO$_x$:H/c-Si (FZ)/a-SiO$_x$:H structure, equally treated on both sides (as seen in Fig. 6.2). If needed, intrinsic amorphous silicon layers a-Si:H(i) are deposited for comparison at 13.56 MHz with identical process parameters, but without CO$_2$.

Initially, the silicon wafers are cleaned using the standard RCA procedure followed by a dip in fluoric acid (HF) prior deposition, as described elsewhere (section A.1). The gas flow rates for SiH$_4$, H$_2$, and CO$_2$ have been varied in the range given by Table 6.2.

Table 6.2.: PECVD conditions for forming a-SiO$_x$:H films on c-Si wafer material at front- and backside. The oxygen content of the a-SiO$_x$:H films are controlled by varying the CO$_2$ partial pressure χ_{ox} = [CO$_2$]/([CO$_2$]+[SiH$_4$]); depending on the experiment it is optimized in the course of this section.

| | PECVD conditions | | | |
process	heater temperature T_{heat} (°C)	precursor gas flow (sccm)	pressure (mTorr)	plasma frequency (MHz)
RCA cleaning HF-dip a-SiO$_x$:H	200 – 400	H$_2$: 0 – 200 SiH$_4$: 0 – 60 CO$_2$: 0 – 80	200 – 1000	13.56 – 110

The film compositions and also the changes in the microscopic structure of the amorphous network upon thermal annealing are studied in this chapter using Raman spectroscopy, secondary ion mass spectroscopy, lifetime measurments, as well as spectroscopic ellipsometry measurements.

To determine the passivation quality of the a-SiO$_x$:H films, laser-induced microwave-detected photo conductance decay (μ-WPCD), as described in section 3.6.1, as well as quasi-steady-state photoconductance measurements (QSSPC), as described in section 3.6.2, provide a contactless measurement of the effective recombination lifetime of free carriers. The QSSPC effective lifetime curves presented in this chapter are composed of multiple curves measured in QSSPC and TPC mode on the same axis. More information about this procedure can be found in section 3.6.

To study the influence of annealing on the surface passivation quality and network bonding structure, the samples are annealed in a diffusion furnace under forming gas atmosphere (10 at.% hydrogen diluted in nitrogen) at temperatures ranging from 100 °C to 500 °C.

The film compositions and also the changes in the microscopic structure of the amorphous network depending on plasma deposition frequencies and thermal annealing are studied by means of μ-Raman spectroscopy (μ-RS) (as described in section 3.3.2), and optical profiling techniques (as described in section 3.2). For all μ-Raman measurements presented in this section, the ex-

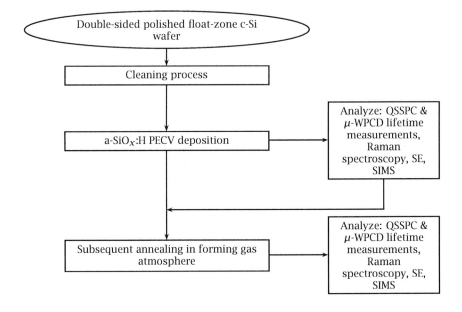

Figure 6.3.: Flowchart illustrating sample preparation of a-SiO$_x$:H passivated c-Si samples.

citation is supplied by an Ar$^+$ ion laser using a wavelength of 488 nm and a power of 25 mW. The spot size of the focused laser light is ~1 μm. The spectra are collected at room temperature by a Peltier cooled CCD detector. The resolution is limited to 0.7 cm^{-1}. The spectra are recorded in the wavenumber range from 200 to 2500 cm^{-1}.

Secondary ion mass spectroscopy gives information about the composition in the microscopic network, in particular the quantity of build-in oxygen and defective carbon atoms. Bandgap analysis of the a-SiO$_x$:H films depending upon used oxygen fraction during deposition is given at the end of this section using spectroscopic ellipsometry. The flowchart in Fig. 6.3 summarizes the experimental processes to analyze the passivation quality of a-SiO$_x$:H.

6.4. Optimization of a-SiO$_x$:H PECV deposition parameters

In this section the PECV deposition parameters for amorphous silicon oxide layers (a-SiO$_x$:H) are studied and optimized in terms of the corresponding surface recombination. Special attention has been paid to the effective lifetime achievable with PECV deposited a-SiO$_x$:H. Investigations on the surface passivation quality of PECV deposited a-SiO$_x$:H films are carried out measuring the effective carrier lifetime using both the μ-WPCD and QSS/TPC techniques.

For the μ-WPCD technique, the effective lifetime has been obtained by mapping the lifetime values on a circular area of 1 cm diameter and calculating the average value. For the QSSPC measurements, the maximum effective lifetime occurred at an injection level (excess carrier density) of approximately $1 \cdot 10^{15}$ cm^{-3}. The simplified formula for a homogeneous, defect-free wafer with a spatially uniform carrier lifetime is given by

$$\frac{1}{\tau_{eff}} = \frac{1}{\tau_{bulk}} + \frac{1}{\tau_{surf}}, \tag{6.1}$$

where τ_{eff} is the measured effective lifetime, τ_{bulk} is the bulk lifetime (combined Auger, radiative, and Shockley-Read-Hall recombination), $\tau_{surface}$ is the characteristic surface recombination lifetime component determined by wafer thickness, d, and surface recombination velocity, S_{eff}. From Equation 6.1 can be concluded, that a better passivation leads to a higher τ_{eff}.

For calculation of S_{eff} it is assumed that both surfaces provide a sufficiently low recombination velocity and have the same values ($S_{eff} = S_{front} = S_{back}$), as the sample structure is symmetric (cf. 6.2). S_{eff} can be deduced by

$$S_{eff} = \frac{W}{2} \left(\frac{1}{\tau_{eff}} - \frac{1}{\tau_{bulk}} \right). \tag{6.2}$$

Evaluation of the surface recombination velocity implies a double-sided c-Si wafer passivation. In this section, double-side polished, 250 μm thick, 0.8-1.2 Ωcm n-type and 0.475-0.525 Ωcm p-type FZ wafers are used for the silicon substrates, as substrates of this type and resistivity are used later on for heterojunction solar cell devices. As the bulk lifetime of each sample is comparable, variations in the measured effective lifetime are due to variations in the surface passivation. Thus, the process parameters that give the highest effective lifetime also give the lowest surface recombination velocity. All silicon substrates are RCA cleaned prior to deposition, including a final etching in diluted HF to remove any oxide formed during RCA clean.

The deposition parameters, as described in the previous section, will be optimized with respect to the following process variables (*cf.* Table 6.2):

- The plasma frequency – varied from 13.56 MHz up to 110 MHz, where 110 MHz is the maximum possible for the system used.

- The plasma power – varied from 2 W to 20 W; subject to a homogeneous deposition.

- The process pressure – varied from 200 mTorr to 1000 mTorr; subject to a homogeneous deposition.

- The dilute carbon dioxide to silane gas flow rate – varied from 0.1 to 80 sccm.

- The dilute hydrogen to silane gas flow rate – varied from 0 - 200 sccm.

- The total silane gas flow rate – varied from 1 - 40 sccm.

- The deposition temperature – varied from $T_{dep} = 110\,^\circ C$ to $T_{dep} = 200\,^\circ C$, where 400 °C is the maximum possible for the system used.

The process parameter remaining is the deposition time, which determines the a-SiO$_x$:H layer thickness. A deposition time of 600 sec is used for the optimization experiments. This corresponds to 100-250 nm thickness of the a-SiO$_x$:H films, depending on the deposition rate.

6.4.1. Effect of plasma frequency

Initially, a-SiO$_x$:H films are deposited at various plasma frequencies in the range of 13.56 MHz up to 110 MHz, and gas compositions containing oxygen fraction of $\chi_{ox} = 0 - 50$ at.%. In this experiment, the intrinsic a-SiO$_x$:H layers are deposited at a heater temperature of 350 °C, corresponding to $T_{dep} = 175\,^\circ C$, and the chamber pressure is set to 300 mTorr. In Fig. 6.4, the effective life-times of samples deposited on p-type FZ wafers with plasma frequencies of 13.56 MHz, 70 MHz and 110 MHz are shown. a-SiO$_x$:H films of two differ-ent compositions are applied: (i) χ_{ox} = 30 % and (ii) χ_{ox} = 50 %. As seen from Fig. 6.4, the highest lifetimes are achieved with a plasma frequency of 70.0 MHz at annealing temperatures of 250 °C. Annealing those samples at higher temperatures leads to an effusion of hydrogen, resulting in a detoria-tion of the surface passivation. The impact of thermal annealing is discussed seperatley and adressed in section 6.5.2.

The effective lifetime of a-SiO$_x$:H passivated c-Si deposited at 110 MHz show an unexpected increase of the effective lifetime at annealing temperatures > 250 °C (up to 500 °C), whereas in contrary the effective lifetime for samples of a-SiO$_x$:H passivated c-Si deposited at lower plasma frequencies drop at that high temperatures around 500 °C. The increase of the effective lifetime for samples deposited with 110 MHz might be due to a micro-crystalline charac-ter of the a-SiO$_x$:H (discussed in section 6.5), preventing a hydrogen effusion during annealing, as well as due to further diffusion of hydrogen to the c-Si interface.

6.4.2. Effect of gas flow rates

In Fig. 6.5 the effective lifetime of a-SiO$_x$:H passivated c-Si samples is displayed as a function of oxygen content (χ_{ox}) in the feedstock during deposition. For this experiment, the deposition temperature has been set to 320 °C and the plasma frequency to 70 MHz. As seen from Fig. 6.5, with increasing χ_{ox} from 0 % to 20 % τ_{eff} increases due to a passivating effect of the oxygen build into the resulting amorphous network during deposition. A highest effective lifetimes of > 2000 μs (corresponding to a low S_{eff} = 6.6 cm/s) is obtained for an oxygen content of χ_{ox} = 20 at.%.

With higher oxygen content than χ_{ox} = 20 at.% in the gas phase, the lifetime decreases. One possible explanation could be that even though the content of carbon contained in the CO_2 is assumed to be below 1 %, the carbon fraction increases with increasing oxygen content, resulting in a higher defect density.

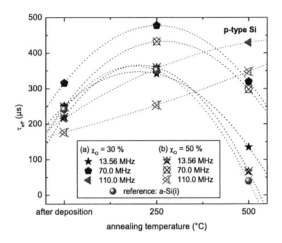

Figure 6.4.: Effective lifetime as a function of annealing temperature depending on the plasma frequency for a-SiO$_x$:H layer deposited at plasma frequencies of 13.56 MHz, 70 MHz and 110 MHz. Two oxygen ratios are applied (a) χ_{ox} = 30 % and (b) χ_{ox} = 50 %. For comparison, the effective lifetime of intrinsic amorphous silicon without addition of oxygen (a-Si:H(i)) is added to the graph. The lines are a guide to the eye. A graph with higher solution for a plasma frequency of 70 MHz is given in Fig. 6.8.

This effect will be investigated further in the following section (*cf*. 6.5), which is dealing with SIMS measurements.

6.4.3. Effect of deposition temperature

A variation of the deposition temperature (also referred to as heater temperature) will be discussed in the following. In subsection 6.4.2, an optimum oxygen content of χ_{ox} = 20 at.% has been found in respect of the obtained effective lifetime. Therefore, for this experiment investigating the deposition temperature dependence, the oxygen content is kept constant at χ_{ox} = 20 at.%. Figure 6.6 shows the effective lifetime of a-SiO$_x$:H passivated c-Si samples as a function of the heater temperature (T_{heat}). The highest effective lifetime is ob-

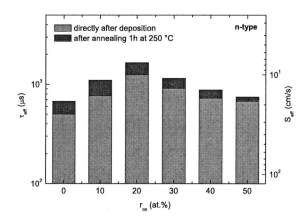

Figure 6.5.: Effective lifetime (τ_{eff}) as a function of oxygen content (χ_{ox}) in the feed stock during the a-SiO$_x$:H decomposition. For each oxygen content, the effect of post-annealing is shown for temperatures of 250 °C and 500 °C.

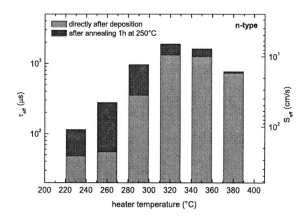

Figure 6.6.: Effective lifetime (τ_{eff}) as a function of the heater temperature (T$_{heat}$) before and after post-anealing at 250 °C. A maximum τ_{eff} is found at a T$_{heat}$ = 320 °C, corresponding to a T$_{dep}$ = 155 °C.

tained for samples deposited at T$_{heat}$ = 320 °C (corresponding to a deposition temperature T$_{dep}$ = 155 °C), as τ_{eff} > 2000 μs, corresponding to S$_{eff}$ < 6.6 cm/s, after subsequent annealing at 250 °C. Thermal annealing at higher temperatures (60 min at 500 °C) leads to a drastic decrease of the effective lifetime for all samples the samples due to the effusion of hydrogen (discussed in section 6.5.2).

6.4.4. Effect of pre-heating time

Initially, the a-SiO$_x$:H films are optimized for a pre-heating time of t$_{heat}$ = 15 min. In this experiment, the PECVD process parameters are kept constant, but the pre-heating time, t$_{heat}$, is varied from t$_{heat}$ = 15 - 120 min in order to study the dependence on temperature changes in the chamber during pre-heating. Figure 6.7 illustrates the dependency of τ_{eff} on the pre-heating time (t$_{heat}$) of the c-Si sample in the PECVD setup for t$_{heat}$ = 15 - 120 min. Comparing this results to the pre-heating curve of our PECVD setup shown in section 3.1.1 (*cf.* Fig. 3.3), it is evident that the pre-heating time becomes a critical factor: a t$_{heat}$ of 15 min results in a T$_{dep}$ of less than 155 °C, hence with increasing t$_{heat}$ time T$_{dep}$ increases. As seen already in Fig. 6.6, T$_{dep}$ needs to be exact 155 °C. The choice of a short pre-heat time and corresponding heater temperature, rather than a longer t$_{heat}$ and lower T$_{heat}$, is purely based on process time within our laboratory.

6.4.5. Impact of post-annealing on passivation properties of a-SiO$_x$:H/c-Si heterostructures

In this experiment, a-SiO$_x$:H films deposited on both 1 Ωcm n-type and 0.5 Ωcm p-type mirror polished FZ-Si substrates, are stepwise annealed in a diffusion oven in a range from 100-500 °C under forming gas atmosphere (10 % H$_2$ in nitrogen) for one hour. For this sample series the plasma frequency and oxygen fraction are set to 70 MHz (corresponding to the highest lifetime values of samples presented in Fig. 6.4) and χ_{ox} = 20 %, respectively. Figure 6.8 shows the changes of the electronic passivation quality due to step-

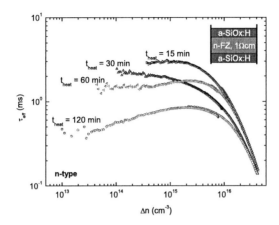

Figure 6.7.: Impact of pre-heat time (t_{heat}) of the c-Si sample in the PECVD setup prior deposition on effective lifetime of a-SiO$_X$:H passivated samples.

wise post-annealing of the a-SiO$_X$:H films deposited on both n-type and p-type c-Si.

For the n-type Si-FZ it is found that subsequent annealing of the samples at 250 °C for 3 h in forming gas atmosphere increases drastically the effective lifetime to its maximum. A high effective lifetime (τ_{eff} = 4.7 ms, leading to $S_{eff} \leq 2$ cm/s) has been achieved for samples deposited with an oxygen content of 20 at.% after subsequent annealing at 250 °C. This might be (i) due to hydrogen saturation of Si dangling bonds at the a-SiO$_X$:H surface, or (ii) due to the generation of Si-(OH)$_x$ and Si-O-Si bonds in the microscopic structure of the amorphous network (*cf.* Raman analysis in section 6.5).

Thermal annealing at higher temperature (500 °C) leads to a steady decrease of the effective lifetime of the samples due to (i) an effusion of hydrogen or (ii) C increases with increasing oxygen content from the carbon dioxide, resulting in a higher defect density. Also, for a-Si:H(i) material annealed at higher temperatures, a correspondence between the H$_2$ effusion rate and defect generation in the film has been demonstrated in the past by comparing TDS, IR absorption, and electron spin resonance measurements (*cf.* [153]). The same

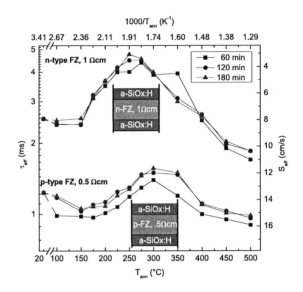

Figure 6.8.: Influence of stepwise annealing T_{ann} step on the electronic passivation quality, expressed by the effective lifetime τ_{eff} and the surface recombination velocity S_{eff} of thin intrinsic a-SiO$_x$:H layers deposited on mirror polished 1 Ωcm n-type FZ-Si and 0.5 Ωcm p-type FZ-Si surfaces. The lines are guides for the eye.

mechanism can be attributed to a-SiO$_x$:H, where hydrogen likely is transferred at temperatures above 350 °C from a Si–H to a H$_2$ state, creating defects in the material (*cf.* section 6.5).

As stated in [143], the quality of a passivation scheme applied to a silicon material of different conduction type differs. This is consistent for the a-SiO$_x$:H passivated p-type Si-FZ substrates: resulting in a lower effective lifetime of τ_{eff} > 2 ms and a shift of the maximum of τ_{eff} towards 300 °C (compared to a-SiO$_x$:H passivated n-type material with a maximum of τ_{eff} around 250 °C annealing temperature). The lower lifetime of p-type passivated material (τ_{eff} > 2 ms) compared to n-type passivated material (τ_{eff} > 4 ms) might be attributed mainly to the defect states in our non-stoichiometric a-SiO$_x$:H films. It can be assumed that some of these defect states are within the band gap of the c-Si at the interface, and therefore directly induce the recombination at the interface to the c-Si material of different conduction type.

6.4.6. Optimization results

In conclusion, the PECVD plasma parameters are optimized in terms of measured effective lifetime. Table 6.3 displays these optimized deposition parameters developed in this section. Hereafter, these parameters are used to prepare high quality a-SiO$_x$:H films, unless stated different.

Table 6.3.: Optimized deposition conditions for the developed a-SiO$_x$:H films.

process parameter	a-SiO$_x$:H deposition conditions
[CO$_2$]:([CO$_2$]+[SiH$_4$])	1:5
pressure (mTorr)	200
deposition rate (Å/s)	3.5
plasma frequency (MHz)	70
T$_{dep}$ (°C)	155
power density (mW/cm^2)	40
electrode distance (mm)	19
post-annealing in forming gas (10 % H$_2$ diluted in N)	250 °C for n-type c-Si 300 °C for p-type c-Si

6.5. Compositional analysis of a-SiO$_x$:H

6.5.1. Changes in the microscopic structure upon plasma frequency

Changes in the microscopic structure upon thermal annealing are investigated using Raman spectroscopy. Figure 6.9 displays the corresponding Raman spectra of films deposited with plasma frequencies at 13.56 MHz, 70 MHz and 110 MHz. In addition, the spectra of the reference a-Si:H(i) layer are superposed as a reference spectrum. As seen from Fig. 6.9, the character of films deposited at 13.56 MHz can be attributed to amorphous, whereas the character of films deposited 70 MHz is close to the transition to micro-crystalline around 500 cm^{-1}. The spectra of the films deposited at 110 MHz exhibit a pronounced peak at 500 cm^{-1} corresponding to a micro-crystalline character.

6.5.2. Changes in the microscopic structure upon thermal annealing

Changes in the microscopic structure upon thermal annealing are investigated using Raman spectroscopy. Figure 6.10 exhibit the Raman spectra for an a-SiO$_x$:H film deposited at 70 MHz, directly after deposition and after subsequent post-annealing at 250 °C.

From Fig. 6.10 can be concluded that after post-annealing of the film, a generation of Si-(OH)$_x$ bonds appear with a peak formation around 920 cm^{-1} and 1079 cm^{-1}, which agrees well with the results shown by *Hoex et al.* [154] and *Kubicki and Sykes* [155]. Furthermore, Si-O-Si bonds occur around 820 cm^{-1}, whereas the reference sample a-Si:H(i) depicted in Fig. 6.11 does not show any compositional changes due to thermal annealing in the microscopic structure. However, the generation of Si-O-Si and Si-(OH)$_x$ bonds due to thermal annealing does not depend on the applied plasma frequency, as can be seen in Fig. 6.11.

For samples annealed at higher temperatures (T$_{ann}$ = 500 °C for 1 h) significantly lower Si-(OH)$_x$ or Si-O-Si peaks are detected (not shown in the figures). Effusion of hydrogen at temperatures around 400 - 500 °C might be the reason, as reported by *Park et al.* [156]. Therefore, one can conclude that plasma frequencies influence only the amorphous/micro-crystalline character during deposition, and show no effect on the generation of bonds like Si-(OH)$_x$ or Si-O-Si after thermal annealing.

Figure 6.11 show the Raman spectra for a-SiO$_x$:H films deposited at 13.56 MHz, 70 MHz and 110 MHz after thermal annealing at 250 °C superposed to a standard a-Si:H(i) film deposited at 13.56 MHz. The same conclusions for samples deposited with plasma frequencies of 13.56 MHz and 110 MHz can be drawn as observed for samples deposited at 70 MHz. For samples deposited at 110 MHz, the generation of Si-O-Si bonds and Si-(OH)$_x$ bonds appears to

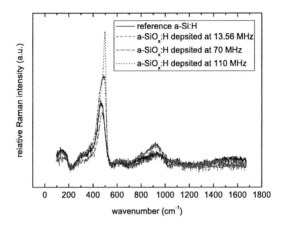

(a) Full range between 200 cm^{-1} to 1600 cm^{-1}.

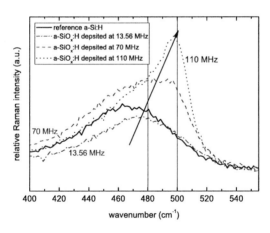

(b) Close-up range between 400 cm^{-1} and 500 cm^{-1}.

Figure 6.9.: Raman spectra of a-SiO$_x$:H(i) films directly after deposition using plasma frequencies of 13.56 MHz, 70 MHz and 110 MHz. For comparison the spectra of intrinsic a-Si:H(i) are superposed.

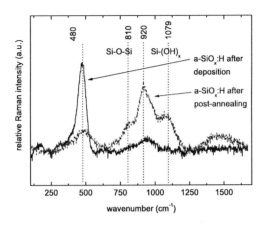

Figure 6.10.: Raman spectra of a-SiO$_x$:H films deposited at 70 MHz: the spectra of a sample directly after deposition and after thermal annealing at 250 °C are superposed. A generation of Si-O-Si bonds as well as Si(OH)$_x$ bonds is observed.

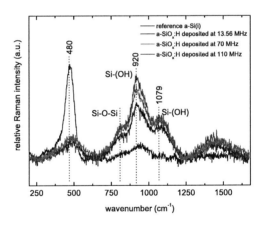

Figure 6.11.: Raman spectra after thermal annealing at 250 °C deposited using plasma frequencies of 13.56 MHz, 70 MHz and 110 MHz. For comparison the spectra of intrinsic a-Si:H(i) after thermal annealing are superposed.

be less pronounced compared to the samples deposited with 13.56 MHz and 70 MHz, because the intensity is typically measured in arbitrary units.

As shown in section 6.4.5, subsequent annealing of the samples drastically increases the effective lifetime. This might be (i) due to hydrogen saturation of Si dangling bonds at the a-SiO$_x$:H(i) surface, or (ii) due to compositional changes in the microscopic structure of the amorphous network. In case of standard a-Si:H(i) ($\chi_{ox} = 0$ at.%), the hydrogen content is roughly 12-14 at.% as reported by [150] and [61]. It is also known that the amorphous silicon sub-oxides contain a significant higher fraction of hydrogen.

Raman spectra around the 2000 cm^{-1} band region show typical peaks for Si-H bondings for all prepared a-SiO$_x$:H samples (not depicted in the Figures). Therefore, it can be assumed that the fraction of hydrogen in the films is sufficiently high to cover all dangling bonds and that no additional hydrogen is build up in the amorphous network from annealing in forming gas. The concentration of carbon (C) and other impurities is below 1 at.% for all samples investigated in this section.

Nevertheless, lattice distortions, such as SiO$_2$ or Si clusters, silicon dangling bonds, molecular hydrogen inclusions, or micro-voids with hydrogen terminated internal surfaces are likely to exist in the a-SiO$_x$:H(i) samples.

6.5.3. Secondary ion mass spectroscopy analysis of a-SiO$_x$:H films

Secondary ion mass spectrometry (SIMS) has been carried out to analyze the composition of the a-SiO$_x$:H films. Explicitly, measurements of films with an oxygen content of $\chi_{ox} = 0$ at.%, $\chi_{ox} = 20$ at.% and $\chi_{ox} = 50$ at.% are compared and depicted in Fig. 6.12. The choice of the χ_{ox} ratio is based on the lifetime results previously achieved, where the sample with $\chi_{ox} = 20$ at.% showed the highest effective lifetime, the sample with $\chi_{ox} = 0$ at.% is used as reference and the sample with $\chi_{ox} = 50$ at.% represent the highest investigated oxygen content.

It is found that oxygen is build into the amorphous network; with increasing oxygen content in the feed stock during deposition, SIMS reveals an increasing oxygen content in the resulting network. A higher oxygen content leads also to an increased deposition rate (from 3.0 Å/s up to 3.8 Å/s), indicated by the

interface peaks. The pile-ups visible at the interface air/a-SiO$_x$:H and at the
interface a-SiO$_x$:H/c-Si can be explained as segregation effects. Furthermore,
the carbon fraction in the a-SiO$_x$:H films increases with higher χ_{ox} content
during deposition. This result explains the decrease of the effective lifetime
at χ_{ox} = 50 at.% (shown in Fig. 6.5); indicating a higher defect density.

6.5.4. Optical confinement of a-SiO$_x$:H layers

From spectroscopic ellipsometry (SE) measurements of the a-SiO$_x$:H films, the
real part of the refractive index n and the extinction coefficient k can be deduced, as described in section 3.2. For this purpose, a-SiO$_x$:H films are deposited on a c-Si/SiO$_2$ substrate; the 1000 Å thick SiO$_2$ layer is then employed
as an optical separation layer to enhance contrast. For the SE analysis, a simple
three layer model c-Si/SiO$_2$/a-SiO$_x$:H is used (cf. section 3.2). The deposited
thickness is determined by profilometer measurements. The absorption coefficient α is then calculated by means of

$$\alpha = \frac{4 \cdot \pi \cdot k}{\lambda}. \tag{6.3}$$

Fig. 6.13 shows α as a function of the photon-energy for the a-SiO$_x$:H(i) films
compared to that of standard a-Si:H(i) and crystalline silicon. The absorption
in the a-SiO$_x$:H film at the blue light region around 3.0 eV is significant lower
than that of a-Si:H(i). Consistently, the fraction of blue light transmitted to
the wafer which is available for solar conversion increases drastically (though
depending on the applied thickness).

As the passivation quality depends strongly on the thickness of an passivating layer [14, 86], it is evident that with decreasing absorption in the passivation layer, the thickness could be increased to gain a better passivation quality.
The surface passivation depending on applied layer thickness of a-SiO$_x$:H will
be investigated in section 6.6.5 in detail.

The optical bandgap E$_{04}$ of the a-SiO$_x$:H films strongly depend on χ_{ox} and linearly rise with increasing oxygen content from 1.7 eV (χ_{ox} = 0 at.%) up to 2.4 eV
(χ_{ox} = 50 at.%), cf. Fig. 6.14. Thermal annealing at 250 °C does not noticeably
influence the resulting bandgap. As has been discussed earlier the annealing

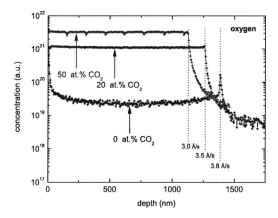

(a) SIMS spectra of samples deposited with different χ_{ox} rates exhibiting the oxygen content in the a-SiO$_x$:H film.

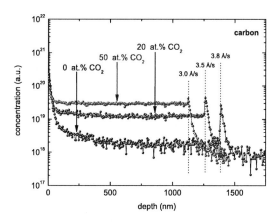

(b) SIMS spectra of samples deposited with different χ_{ox} rates exhibiting the carbon content in the a-SiO$_x$:H film. The concentration of carbon appears to be less than 1 vol. %.

Figure 6.12.: Secondary ion mass spectrometry measurements of a-SiO$_x$:H layers deposited with oxygen contents in the feed stock during deposition of χ_{ox} = 0 at.%, χ_{ox} = 20 at.%, and χ_{ox} = 50 at.%. The film thicknesses are estimated by profilometry.

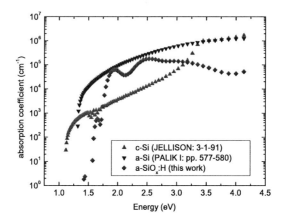

Figure 6.13.: Absorption coefficient α of a-SiO$_X$:H deduced from SE data fit compared to published absorption coefficients of crystalline silicon (*cf*. Jellison 3-1-91) and a-Si:H (*cf*. Palik HOC I, 577-580). The absorption coefficient at the blue light region around 3.0 eV is by one order of magnitude lower than the one of standard a-Si:H.

at higher temperatures (500 °C) leads to a decrease of E$_G$ (not depicted), which can be as well attributed to the effusion of hydrogen. The optical band gaps of samples deposited with 110 MHz are constant even after annealing at higher temperatures (500 °C), due to the micro-crystalline structure.

6.6. Surface passivation quality of a-SiO$_x$:H

In the previous section, the optimization of the PECV deposition parameters of the a-SiO$_x$:H layers (*cf*. section 6.4) is described. This section refers to the surface passivation quality of a-SiO$_x$:H compared to passivation quality of various passivation schemes published elsewhere, among those the standard SiO$_2$, SiN$_x$ and a-Si:H(i).

Furthermore, this section contains a study of a-SiO$_x$:H passivating c-Si of various doping levels and types. The homogeneous distribution of the measured τ_{eff} will be investigated, as well as the dependency of the applied

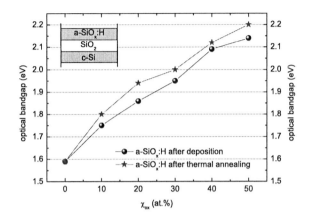

Figure 6.14.: Optical bandgap of a-SiO$_x$:H films deposited at 70 MHz as a function of χ_{ox}; after deposition and after post-annealing at 250 °C. The sample configuration used for the bandgap determination is displayed in the inlet.

a-SiO$_x$:H layer thickness in respect of the application to heterojunction solar cell devices in stacks with a-Si:H films are compared.

Comparing different passivation schemes such as SiO$_2$ and SiN$_x$ published elsewhere implies identically measurement conditions as well as identically analysis of the measured data. Passivation quality depends on applied film thickness of the passivating film, and the doping type and level of the material used.

Deciding on the specified mean carrier density (MCD) specifying a value for τ_{eff} is therefore discretionary. Ideally, a MCD is chosen that corresponds to the V_{mp} in the used material. However, $1 \cdot 10^{15}$ cm^{-3} is very appealing, because it is well above the noise, above trapping and above DRM effects. Some authors tend to use $5 \cdot 10^{14}$ cm^{-3}, some prefer $1 \cdot 10^{15}$ cm^{-3}. Reporting both the J$_0$ and the τ_{eff} give the opportunity to compare data at both $5 \cdot 10^{14}$ cm^{-3} to $1 \cdot 10^{15}$ cm^{-3}. To ensure comparability, a MCD of $1 \cdot 10^{15}$ cm^{-3} is used to specify the value of τ_{eff} hereafter, unless stated different.

As the sample structure is symmetrical (*cf.* Fig. 6.2), all S$_{eff}$ values presented hereafter are calculated under the assumption that both surfaces provide a sufficiently low recombination velocity and have the same values ($S_{eff} = S_{front} = S_{back}$). S_{eff} is related to τ_{eff} by

$$\frac{1}{\tau_{eff}} = \frac{1}{\tau_{bulk}} + \frac{2S_{eff}}{W}, \tag{6.4}$$

so S_{eff} is deduced by

$$S_{eff} = \frac{W}{2}\left(\frac{1}{\tau_{eff}} - \frac{1}{\tau_{bulk}}\right). \tag{6.5}$$

One should note that the uncertainty of S_{eff} depends strongly on the value used for τ_{bulk}, as described in detail in section 2.1. Therefore, an upper limit of the surface recombination velocity (SRV) is calculated for the case that (i) no Shockley-Read-Hall (SRH) recombination is considered (by setting $\tau_{bulk} \rightarrow \infty$ in equation 6.1, leading to an approximate value of $S_{eff} = d \cdot (2 \cdot \tau_{eff}{}^{-1})$, with d defining the wafer thickness.) and (ii) for $\tau_{bulk} = 35$ ms after *Yablonovitch and Gmitter* [157], which is well above the highest ever measured effective lifetime, *cf.* section 6.2.

6.6.1. Mapping of a-SiO$_x$:H surface passivation quality

The homogeneous distribution of the measured τ_{eff} will be investigated in this experiment. Therefore, the μ-WPCD lifetime method allows a mapping of the passivation quality distributed over a passivated sample by taking discrete values of τ_{eff} with a predefined step rate and sample size. Fig. 6.15 illustrates the surface passivation quality distribution of a-SiO$_x$:H films deposited on both sides of an 1 Ωcm n-type FZ material before and after annealing of those films. From the results shown in Fig. 6.15 it is evident that a passivation quality exceeding 4 ms after annealing is distributed homogeneously over the whole passivated c-Si wafer. The average τ_{eff} value is found to be 4075 μs. The black points are errors in measurement.

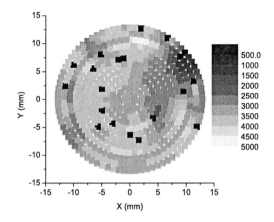

(a) τ_{eff} mapping of a-SiO$_x$:H passivated c-Si 1 Ωcm n-type FZ material before annealing of a-SiO$_x$:H.

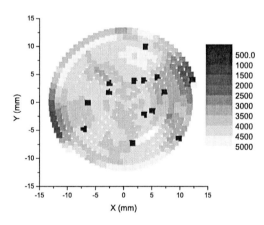

(b) τ_{eff} mapping of a-SiO$_x$:H passivated c-Si 1 Ωcm n-type FZ material after annealing of a-SiO$_x$:H.

Figure 6.15.: μ-WPCD mapping measurement results of τ_{eff}. The scale on the right
shows τ_{eff} in (μs).

6.6.2. Surface passivation quality of a-SiO$_x$:H compared to standard a-Si:H(i)

As for heterojunction solar cells, a-Si:H(i) is the standard material applicable as buffer layer between c-Si and doped a-Si:H, as well as surface passivation of the c-Si material. Therefore, the surface passivation quality of a-SiO$_x$:H is compared to the passivation quality of standard a-Si:H(i) in this experiment. The passivation quality of a-SiO$_x$:H compared to a-Si:H(i) is characterized by the effective lifetime on 1 Ωcm n-type FZ material. Both a-SiO$_x$:H and a-Si:H(i) are deposited on front- and backside with equal process conditions, as given in Table 6.4.

Table 6.4.: Deposition conditions for a-SiO$_x$:H and a-Si:H(i) films.

process parameter	a-Si:H(i)	a-SiO$_x$:H
[CO$_2$]:([CO$_2$]+[SiH$_4$])	-	1:5
pressure (mTorr)	200	200
deposition rate (Å/s)	3.0	3.5
plasma frequency (MHz)	70	70
T$_{dep}$ (°C)	155	155
power density (mW/cm^2)	40	40
electrode distance (mm)	20	19

As illustrated in Fig. 6.16, an excellent effective lifetime of 4.1 ms directly after deposition of a-SiO$_x$:H on a 1 Ωcm n-type FZ wafer has been achieved. After post-annealing at 250 °C for a duration of 3 h in hydrogenated atmosphere the τ_{eff} increased to 4.7 ms. This is by far the highest ever measured effective lifetime value for a PECV deposited amorphous silicon passivated, high-doped, 1 Ωcm n-type wafer. This enhancement for the effective lifetime compared to previous results can be attributed to an additional chamber cleaning and a 10 min dummy process prior deposition of the a-SiO$_x$:H films.

In contrary, effective lifetime values for the best a-Si:H(i) films as deposited with the deposition parameters given in Table 6.4, are 1.7 ms after deposition and 2.1 ms after post-annealing at 250 °C. The results for the a-Si:H(i) films passivating the 1 Ωcm n-type wafer are consistent with other publications. Table 6.5 summarizes the obtained values for τ_{eff} and S_{eff}.

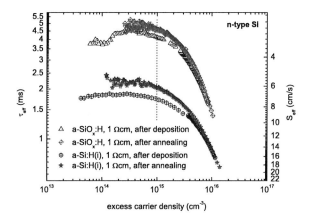

Figure 6.16.: Measured effective lifetime for a-SiO$_x$:H(i) passivated n-type FZ wafers and standard intrinsic a-Si:H(i) passivated n-type FZ wafers as a function of injection level. After additional chamber cleaning, the value for τ_{eff} increased up to 4.7 ms (at $1 \cdot 10^{15}$ cm^{-3}) for the a-SiO$_x$:H on 1 Ωcm n-type FZ. In comparison, for the best a-Si:H(i) layer, a value for τ_{eff} of 2.1 ms has been obtained. The sample structure for measuring with QSSPC is given in Fig. 6.2.

6.6.3. Surface passivation quality of a-SiO$_x$:H compared to standard SiO$_2$ and SiN$_x$ passivation schemes

As described in section 6.2, standard passivation schemes include high quality SiO$_2$ and SiN$_x$ films for high efficiency device fabrication. Record values are established by *Kerr and Cuevas* published in [26], reporting the highest effective lifetime of τ_{eff} = 29 ms, corresponding to the lowest S$_{eff}$ with 0.46 cm/s for a 90 Ωcm FZ *n*-type material passivated with annealed SiO$_2$.

Note, that these values are gained using the effective lifetime at its maximum, disregarding the injection level. However, the SRV for higher doped material, i.e., 1.5 Ωcm and 0.6 Ωcm material, increased to 2.4 cm/s and 26.8 cm/s, respectively, for wafers passivated with SiO$_2$, as shown in Fig. 6.1.

Comparing those results to our passivated 1 Ωcm *n*-type FZ with a-SiO$_x$:H, a SRV of outstanding 2.6 cm/s using a PECVD grown a-SiO$_x$:H passivation

Table 6.5.: Results for passivated c-Si wafers, depending on doping level and applied passivation mechanism - maximum measured effective lifetimes, τ_{eff}, at $1 \cdot 10^{15}$ cm^{-3}, and the corresponding surface recombination velocity (SRV) S$_{eff\text{-}max}$ (for an upper limit by assuming ($\tau_{bulk} \rightarrow \infty$)) and S$_{eff}$ (for τ_{bulk} = 35 ms).

dopant type	doping level (Ωcm)	doping concentration (cm^{-3})	passivation mechanism	τ_{eff} @ $1 \cdot 10^{15}$ cm^{-3} (ms)	S$_{eff\text{-}max}$ $\tau_{bulk} \rightarrow \infty$ (cm/s)	S$_{eff}$ τ_{bulk} = 35 ms (cm/s)
n	1	$5 \cdot 10^{15}$	a-Si:H	1.7	7.35	6.99
n	1	$5 \cdot 10^{15}$	a-Si:H, annealed	2.1	5.95	5.59
n	1	$5 \cdot 10^{15}$	a-SiO$_x$:H	4.1	3.05	2.69
n	1	$5 \cdot 10^{15}$	a-SiO$_x$:H, annealed	4.7	2.65	2.30

scheme is reached (assuming that $\tau_{bulk} \rightarrow \infty$), see Fig. 6.17b. In Fig. 6.17, this record SiO$_2$ passivation of *Kerr and Cuevas* is superposed to the QSSPC results of a-SiO$_x$:H on 1 Ωcm FZ material. An excellent τ_{eff} of > 4 ms on 1 Ωcm material is repeatedly reached and confirmed by different QSSPC setups. This result appears to be the highest measured effective lifetime for 1 Ωcm passivated c-Si material.

6.6.4. Surface passivation quality for bulk material of different doping type and level

The most research publications dealing with surface passivation schemes use low-doped substrates, stating that those wafers contain less defect states, although a subsequent solar cell device is fabricated on high-doped wafer around 1 Ωcm.

The results achieved in this work with a-SiO$_x$:H as a high quality surface passivation scheme are compared to other passivation schemes, by applying a-SiO$_x$:H films to lower doped wafer substrates. The effective lifetime measured by QSSPC and TPC depends strongly on the doping level and type used, as seen i.e., [26, 27] (*cf.* Fig. 6.1). A selection of a 130 Ωcm p-type FZ wafer,

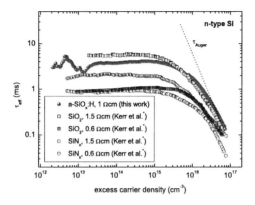

(a) Measured effective lifetime, τ_{eff}, for a-SiO$_x$:H passivated n-type FZ wafers as a function of the excess carrier density in the range of 10^{12}-10^{17} cm^{-3}. Auger recombination dominates at the higher carrier injection levels.

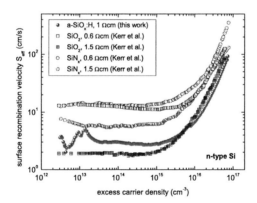

(b) Calculated surface recombination velocity (SRV), S$_{eff}$, as a function of the excess carrier density for n-type FZ silicon material passivated with a-SiO$_x$:H films. Low SRV values of 2.6 cm/s have been achieved for a resistivity of 1 Ωcm.

Figure 6.17.: n-type silicon material passivated with a-SiO$_x$:H. Previously reported passivation schemes of SiO$_2$ and SiN$_x$ films by *Kerr and Cuevas* published in [26, 27] are superposed.

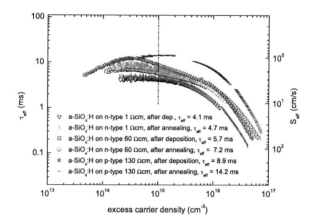

Figure 6.18.: Measured effective lifetime for a-SiO$_X$:H passivated *n*-type and *p*-type FZ wafers of various doping concentrations as a function of injection level. The sample structure for measuring with QSSPC is given in Fig. 6.2.

as well as 60 Ωcm and 1 Ωcm n-type FZ wafers, which have been passivated using a-SiO$_x$:H with subsequent annealing, are presented.

The choice of FZ material with 60 and 130 Ωcm resistivity is purely based on wafer availability within our laboratory, as those wafers are sourced from commercial stock. The results of the measured τ_{eff} and corresponding S_{eff} are presented in Fig. 6.18. A maximum effective lifetime of 14.2 ms has been measured for lightly doped *p*-type material at an injection level of $1 \cdot 10^{15}$ cm^{-3}. This is by far the highest effective lifetime ever measured for a PECV deposited amorphous silicon passivated wafer. In Table 6.6 the results of our passivation scheme are listed in detail.

In Fig. 6.19 the surface passivation quality of a-SiO$_x$:H on low-doped c-Si wafers is compared once more to the best results of previously reported results for passivation schemes using SiO$_2$ and SiN$_x$ films by *Kerr and Cuevas*. It is obvious that the surface passivation quality of a-SiO$_x$:H can be ascertained as good or slightly better than commonly used SiO$_2$ or SiN$_x$.

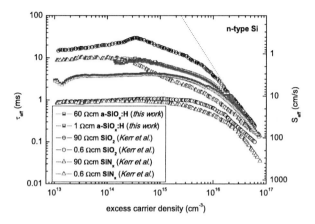

(a) Measured effective lifetime for a-SiO$_x$:H passivated *n*-type FZ wafers.

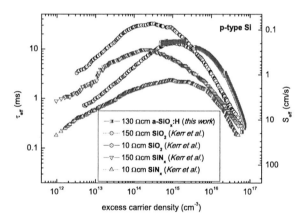

(b) Measured effective lifetime for a-SiO$_x$:H passivated *p*-type FZ wafers.

Figure 6.19.: Measured effective lifetime for a-SiO$_x$:H passivated c-Si FZ wafers of various doping concentrations as a function of injection level. Previously reported results for passivation schemes of SiO$_2$ and SiN$_x$ films by *Kerr and Cuevas* published in [26, 27] are superposed.

Table 6.6.: Results for passivated silicon wafers, depending on doping type and doping level - maximum measured effective lifetimes, τ_{eff}, at $1 \cdot 10^{15}$ cm^{-3}, and the corresponding surface recombination velocity (SRV) S$_{eff\text{-}max}$ (for an upper limit by assuming ($\tau_{bulk} \rightarrow \infty$)) and S$_{eff}$ (for τ_{bulk} = 35 ms).

dopant type	doping level (Ωcm)	doping concentration (cm^{-3})	passivation mechanism	τ_{eff} @ $1 \cdot 10^{15}$ cm^{-3} (ms)	S$_{eff-max}$ $\tau_{bulk} \rightarrow \infty$ (cm/s)	S$_{eff}$ τ_{bulk} = 35 ms (cm/s)
n	1	$5 \cdot 10^{15}$	a-SiO$_x$:H	4.7	2.66	2.30
n	60	$7 \cdot 10^{13}$	a-SiO$_x$:H	7.2	1.73	1.37
p	130	$1 \cdot 10^{14}$	a-SiO$_x$:H	14.2	0.88	0.52

6.6.5. Surface passivation quality depending on applied a-SiO$_x$:H film thickness

Here, the surface passivation quality of a-SiO$_x$:H films depending on the applied film thickness is investigated. For better comparability, a 1 Ωcm n-type FZ wafer is used for all depositions. The thicknesses of the deposited films vary from 250 nm down to 2 nm, which is the lowest thickness ensuring a homogeneous, shunt-free layer distribution. The results of the effective lifetime τ_{eff} and corresponding S$_{eff}$ values for an a-SiO$_x$:H passivated 1 Ωcm n-type wafer are shown in Fig. 6.20. As seen in Fig. 6.20, a film thickness as low as 10 nm (deposition time 35 sec at a rate of 3 Å/s) provides an almost as good surface passivation quality as a 250 nm thick layer. Further reduction of the thickness down to 3 nm results in a slight decrease of the effective lifetime to 3.1 ms. For a film thickness lower than 2 nm the a-SiO$_x$:H films are not deposited homogeneously onto the wafer surface in the PECVD setup. Therefore, the effective lifetime decreases drastically and those results are neglected in the Fig. 6.20. To apply the optimal a-SiO$_x$:H film layer as passivation buffer layer sandwiched between c-Si and doped a-Si:H in heterojunction solar cells, the film thickness is determined by a trade-off between passivation quality, current losses due to low conductivity of the a-SiO$_x$:H films resulting in high series resistances, light absorption and deposition homogeneity, *cf.* section 7.3.3.

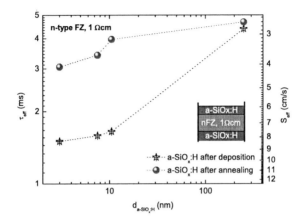

Figure 6.20.: Values for τ_{eff} and S_{eff} as a function of applied a-SiO$_x$:H film thickness, d$_{a\text{-SiO}_x\text{:H}}$, extracted from QSSPC and TPC measurements at an excess carrier density of $1 \cdot 10^{15}$ cm^{-3}. For comparison, films directly after deposition and after annealing are superimposed. The lines are a guide to the eye.

6.6.6. Surface passivation quality of stacked a-SiO$_x$:H(i) and doped μc-Si:H (n$^+$/p$^+$) films

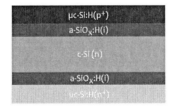

Figure 6.21.: Cross-sectional illustration of the sample configuration.

To determine the passivation quality of our a-SiO$_x$:H films in stacks with the doped μc-Si:H(p) and μc-Si:H(n) films described in the previous chapter (*cf.* chapter 5), the effective lifetime of 1 Ωcm n-type wafers passivated symmetrically with a-SiO$_x$:H, doped μc-Si:H(p) emitter on the topside and doped

μc-Si:H(n) BSF on the backside is measured. Figure 6.21 illustrates the sample configuration. Earlier works (e.g., *Wolf and Beaucarne* [142]) state that whereas

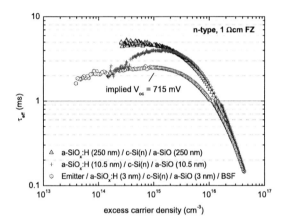

Figure 6.22.: Measured effective lifetime *vs.* injection level (excess carrier density) of substrates passivated with a-SiO$_X$:H films compared to substrates passivated with a-SiO$_X$:H and covered additionally with a-Si:H(p) on the frontside and a-Si:H(n) on the backside. The thickness of the a-SiO$_X$:H layer is well optimized for heterojunction solar cell device fabrication.

even a very thin layer of a-Si:H(i) around 3 nm may still yield a very low surface recombination velocity on low resistivity wafers (0.5 - 1.5 Ωcm), the surface passivation properties are lost when the intrinsic film is subsequently covered by a PECVD a-Si:H(p) layer.

Nevertheless, in experiments performed in this work, the effective lifetime of an 1 Ωcm *n*-type wafer covered by symmetrical 3 nm a-SiO$_X$:H films (thickness is well optimized for heterojunction solar cell device fabrication) plus additional emitter- and BSF-layer, drops only slightly from 4 ms down to 2.5 ms. The surface recombination is hereby adequately prevented. These results are shown in Fig. 6.22, illustrating QSSPC curves for a 1 Ωcm *n*-type c-Si substrate passivated with a 3 nm thick a-SiO$_X$:H film plus an additional doped emitter and BSF.

As references, the QSSPC curves for a 1 Ωcm *n*-type c-Si substrate passivated with (i) a 250 nm and (ii) a 10.5 nm a-SiO$_x$:H film are superposed. An implied open-circuit voltage of 715 mV can be read from this structure. For the case of a-SiO$_x$:H covered by the doped emitter and BSF, a detailed study on the influence of a-SiO$_x$:H film thickness on the heterojunction solar cell performance is given in the next chapter (*cf*. section 7.3.3).

The developed and optimized a-SiO$_x$:H passivation scheme will be further investigated as part of a heterojunction solar cell device described in chapter 7.

6.7. Chapter summary

A novel passivation scheme, featuring PECV deposited amorphous silicon sub-oxide films, has been extensively investigated. The a-SiO$_x$:H films are deposited by decomposition of silane (SiH$_4$), carbon dioxide (CO$_2$) and hydrogen (H$_2$) as source gases. The plasma deposition parameters of a-SiO$_x$:H films were optimized in terms of effective lifetime, whereas the oxygen content and the resulting optical band gap E$_{04}$ of the a-SiO$_x$:H films were controlled by varying the CO$_2$ partial pressure [CO$_2$]/([CO$_2$]+[SiH$_4$]). An optimum substrate deposition temperature and plasma frequency of T$_{dep}$ = 155 °C and 70 MHz, respectively, have been revealed. Also, an optimized gas flow ration has been employed as χ_{ox} = [CO$_2$]:([CO$_2$]+[SiH$_4$]) = 20 at.%. Additional hydrogen reduces the oxygen content to a total of 10 at.% in the gas flow during deposition.

Record high effective lifetime values of 4.7 ms on 1 Ωcm n-type FZ wafers and 14.2 ms on 130 Ωcm p-type FZ wafers prove the surface passivation applicability of silicon wafers to any kind of silicon based solar cells. The values achieved in this work appear to be the highest ever reported values of τ_{eff} on a high doped crystalline wafer of 1 Ωcm resistivity. Additionally, the a-SiO$_x$:H films yield a surface passivation quality exceeding earlier published record passivation schemes such as SiN$_x$ and SiO$_2$. Therefore, the use of a-SiO$_x$:H may be a promising alternative for any passivation scheme existing so far. Advantages of the a-SiO$_x$:H passivation scheme are that the fabricated a-SiO$_x$:H layers are grown by simple PECV deposition at low temperatures, they withstand hydrofluoric acid (tested with a 5 % diluted HF dip for 1 min) and high temperatures up to 500 °C (tested for 1 hour under nitrogen atmosphere). It has to be mentioned that the process parameters have to be carefully adjusted to obtain such high quality a-SiO$_x$:H films.

A high-transparent window layer up to 2.4 eV, depending on the oxygen fraction in the precursor gas, is obtained by adding oxygen to the a-SiO$_x$:H layers. The a-SiO$_x$:H films exhibit a very low blue light absorption compared to a-Si:H(i), enabling a thicker passivation layer than standard a-Si:H(i) and therefore an increasing passivation quality.

The impact of post-annealing at low temperatures around 250 °C of the a-SiO$_x$:H films showed a beneficial effect in form of a significant increase of the effective lifetime. The improvement of the passivation quality driven by low temperature annealing of the a-SiO$_x$:H likely originates from defect reduction of the film close to the interface. Raman spectra reveal the existence of Si-(OH)$_x$ and Si-O-Si bonds after thermal annealing of the layers, leading to a higher effective lifetime, as it reduces the defect absorption of the sub oxides. Thermal annealing at higher temperature (above 450 °C) results in a decrease of lifetime and to hydrogen effusion.

Furthermore, the applicability of a-SiO$_x$:H layers as a high quality passivation scheme used in heterojunction solar cell fabrication is an effective alternative to standard a-Si:H(i) buffer layers. It is found that the surface passivation quality depends on the deposited layer thicknesses. Particular importance of that parameter becomes evident for the application of those films in the production of heterojunction solar cells. The key factors to obtain high quality films a-SiO$_x$:H films comprehend: (i) the chamber contamination, which depend on previous processes with various precursor gases, (ii) dummy process to cover the reactor walls with a-SiO$_x$:H, (iii) the deposition temperature and the pre-heating time in the chamber, and (iv) the electrode distance.

CHAPTER 7.

HETEROJUNCTION SOLAR CELL STRUCTURES PASSIVATED USING WIDE-GAP PECVD A-SIO$_X$:H

7.1. Introduction

The requirements of high optical depth and perfect charge collection imply very high demands of material quality. Thus, in heterojunction solar cell devices, a-Si:H has been accepted as a suitable heterojunction material forming both emitter and surface passivation layer. Nevertheless, the light absorption in the a-Si:H(i) and the doped a-Si:H layers is rather strong, and the short-circuit current density in the a-Si:H/c-Si solar cells decreases rapidly with increasing a-Si:H layer thicknesses [12]. To improve the short circuit density in a-Si:H/c-Si solar cells further, it is preferable to employ an a-Si:H based alloy that has a larger optical bandgap than standard a-Si:H.

Another aspect in heterojunction solar cell applications is the surface passivation of the crystalline wafer, since the passivation quality is directly related to the open-circuit voltage of the cell. It is well known that for i.e., thin film solar cells the PECVD conditions required for production of high quality a-Si:H are very close to the region where micro-crystalline Si is deposited. However, for the growth of a passivation layer on c-Si these conditions often result in epitaxial growth at the interface, reducing the effective lifetime and the V_{oc} of those devices [134]. In this chapter we combine both the strictly amorphous character of high quality surface passivating a-SiO$_x$:H to ensure an abrupt interface and the high quality (in respect of conductivity and transmission) μc-Si:H layers. For this purpose the combination of wide-gap high transparent hydrogenated amorphous silicon oxide a-SiO$_x$:H layers for high quality surface passivation, with wide-gap high-conductive μc-Si:H layers for use as emitter and BSF in heterojunction solar cell applications, is used. Whereas the

a-SiO$_x$:H feature a high quality surface passivation (suppression of surface recombination), a high transparency and assure an abrupt interface to the c-Si
surface (*cf.* chapter 6), the μc-Si:H layers are introduced on account of high
conductivity and high transparency to suppress absorption. The passivation
potential of the a-SiO$_x$:H films for high quality passivation is already ascertained in chapter 6. The a-SiO$_x$:H films are fabricated by VHF (70 MHz) PECVD
decomposition using SiH$_4$, H$_2$, and CO$_2$ at the low deposition temperature of
155 °C. As described in chapter 5, the μc-Si:H films are formed by VHF (110
MHz) PECVD decomposition of SiH$_4$, H$_2$, and TMB or PH$_3$ for p-type and n-type
films, respectively.

To validate the capability of the intrinsic a-SiO$_x$:H and doped
μc-Si:H films separately, heterojunction solar cells consisting of (front
to back) μc-Si:H(p)/a-Si:H(i)/c-Si(n)/a-Si:H(i)/μc-Si:H(n) as a reference,
μc-Si:H(p)/a-SiO$_x$:H(i)/c-Si(n)/a-SiO$_x$:H(i)/μc-Si:H(n), and the reversed doping sequence have been analyzed. Furthermore, a variation of the applied
film thickness of the grown a-SiO$_x$:H films determines the impact on the
performance of heterojunction solar cells. It is expected that by applying
a-SiO$_x$:H layers to the heterojunction cell devices, the conversion efficiency
can be improved due to higher V$_{oc}$ and FF; and more importantly, Si epitaxial
growth observed at high process temperatures is suppressed completely.

Since the growth of both a-SiO$_x$:H and μc-Si:H films will be influenced by the
wafer surface conditions, both double-sided polished float-zone wafer as well
as textured float-zone wafers are used as substrates. Using this approaches,
the impact of the growth conditions of a-SiO$_x$:H and μc-Si:H on different wafer
substrates, the applicability of the a-SiO$_x$:H as high quality surface passivation,
and the use of wide-gap high conductive μc-Si:H layers as emitter and BSF is
distinguished and allocated.

7.2. Experimental

Figure 7.1 illustrates a cross-sectional view of the heterojunction solar cell
design in this chapter. In order to achieve a higher response in the short-
wavelength region, wide-gap high conductive micro-crystalline emitter films

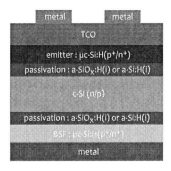

Figure 7.1.: Cross-sectional illustration of applied heterojunction cell design.

are grown. A high hydrogen dilution increases the optical bandgap of the emitter up to 2.0 eV. The micro-crystalline network, however, harms the interface if deposited directly onto the crystalline wafer. An abrupt, in particular an amorphous interface, is necessary to obtain high efficient heterojunction cell devices; therefore, no direct comparison to devices without passivation layer can be done. Hydrogenated amorphous silicon suboxides (a-SiO$_x$:H(i)) films are used as wide-gap high quality surface passivating buffer layer, whereas standard hydrogenated amorphous silicon (a-Si:H(i)) is used as reference. For n-type silicon FZ material, the μc-Si:H(p) and μc-Si:H(n) films are deposited as emitter and back-surface-field, respectively. When p-type c-Si substrates are applied, the heterojunction is formed by μc-Si:H(p)/a-SiO$_x$:H(i)/c-Si(n)/a-SiO$_x$:H(i)/μc-Si:H(n) (from front to back). The various cell designs are listed below, the application of a-Si:H(i) serves as a reference:

- μc-Si:H(p)/a-Si:H(i)/c-Si(n)/a-Si:H(i)/μc-Si:H(n)

- μc-Si:H(p)/a-SiO$_x$:H(i)/c-Si(n)/a-SiO$_x$:H(i)/μc-Si:H(n)

- μc-Si:H(n)/a-Si:H(i)/c-Si(p)/a-Si:H(i)/μc-Si:H(p)

- μc-Si:H(n)/a-SiO$_x$:H(i)/c-Si(p)/a-SiO$_x$:H(i)/μc-Si:H(p)

The process of solar cell fabrication is presented in the flowchart presented in Fig. 7.2. For fabrication of the solar cells, specular float-zone (FZ) silicon wafers with ⟨100⟩ orientation are used. Typically, standard RCA cleaning and termination of the surface by hydrogen using HF solution is applied. The resistivity of the c-Si substrates varies for different materials, so for p-type c-Si based heterojunction devices, p-type wafers of 0.5 Ωcm, and 1 Ωcm are used; whereas for n-type c-Si based heterojunction devices, n-type wafers of 0.5 Ωcm, 1 Ωcm, and 2 - 5 Ωcm are applied. The thickness of the FZ wafers is 250 μm for all samples discussed in this chapter. The TCO, as well as the front- and rear side-electrodes processes are described elsewhere, $cf.$ appendix A. The process parameters forming both a-SiO$_x$:H and μc-Si:H (n$^+$ and p$^+$) are kept constant during the investigations in this chapter, their optimization and corresponding properties are described in earlier chapters (refer to chapter 5 for investigations on μc-Si:H, and to chapter 6 for investigations on a-SiO$_x$:H). Table 7.1 displays the optimized PECVD process parameters for all following investigations.

Table 7.1.: PECVD conditions used for n-type and p-type c-Si based heterojunction solar cells. The dopant sources are PH$_3$ (3 % in SiH$_4$) and TMB (2 % in H$_2$) for n$^+$ and p$^+$ doping, respectively. The gas concentration corresponds to χ_{ox} = [CO$_2$]/([CO$_2$]+[SiH$_4$]), χ_H = [H$_2$]/([H$_2$]+[SiH$_4$]), χ_{PH_3} = [PH$_3$]/([PH$_3$]+[SiH$_4$]), and χ_{TMB} = [TMB]/([TMB]+[SiH$_4$]). For passivation a-SiO$_x$:H(i) is used, the parameters for a-Si:H(i) are used as reference.

	PECVD conditions			
process	a-SiO$_x$:H(i)	a-Si:H(i)	μc-Si:H(p)	μc-Si:H(n)
T$_{dep}$ (°C)	155	155	140	155
precursor gas rate χ_H (%)	50	50	99.01	98.07
precursor gas rate χ_{ox} (%)	20	–	–	–
precursor gas rate χ_{TMB} (%)	–	–	2.62	–
precursor gas rate χ_{PH_3} (%)	–	–	–	1.5
deposition pressure (mTorr)	200	200	250	350
plasma frequency (MHz)	70	70	110	110
plasma power density (mW/cm^2)	34.9	34.9	69.9	69.9
deposition rate (Å/s)	3.5	3.0	1.4	1.4
electrode distance (mm)	19	12	19	19
post-annealing (°C)	250	250	180	180

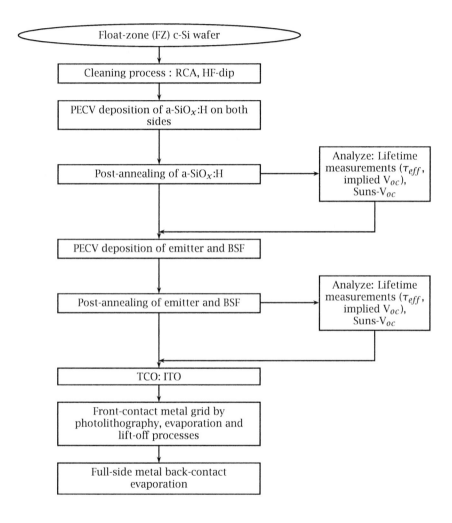

Figure 7.2.: Flowchart illustrating the sample preparation of heterojunction solar cell devices.

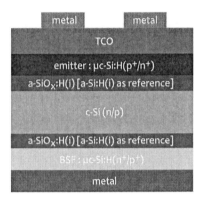

Figure 7.3.: Cross-sectional illustration of applied heterojunction cell design.

7.3. Application of a-SiO$_x$:H and μc-Si:H films to heterojunction solar cells using polished float-zone c-Si

7.3.1. Comparison of a-SiO$_x$:H(i) and a-Si:H(i) used as passivating buffer layer

In heterojunction solar cell applications, a-Si:H(i) has attracted the photovoltaic community due to the success of HIT cells (*SANYO*) [8, 9, 82]. The a-Si:H(i) films are grown by PECV-deposition at low temperatures (<200 °C). However, *Fujiwara and Kondo* [18] stated that the growth of a-Si:H at temperatures T$_{dep}$ > 130 °C often leads to an epitaxial layer formation on the c-Si, reducing the solar cell performance. This undesired presence of an epi-Si layer may however be resolved by a-SiO$_x$:H film deposition, which is grown amorphous with the optimized process parameters shown in section 6.4. Also, due to the inherent strong blue light absorption, only ultra-thin a-Si:H(i) films can be allowed to prevent losses. In this section, standard a-Si:H(i) is compared to a-SiO$_x$:H for high quality surface passivation in respect of the heterojunction solar cell performance. Both *n*- and *p*-type wafers are used as substrates. Figure 7.3 illustrates a cross sectional view of the heterojunction design applied. Table 7.2 summarizes the results for photovoltaic output parameters of heterojunction solar cells passivated with both a-SiO$_x$:H(i) or a-Si:H(i) for

n-type and p-type bulk material. For n-type based heterojunction cells passivated with a-SiO$_x$:H ($\tau_{eff} > 4$ ms, $cf.$ section 6.6) a very high open circuit voltage of 695 mV is found, whereas for cells passivated with standard a-Si:H(i) ($\tau_{eff} > 2$ ms, $cf.$ section 6.6) the open circuit voltage (V_{oc}) reaches only 618 mV. All other parameters are kept constant. One possible reason for the discrepancy in V_{oc} might be the superior surface passivation of a-SiO$_x$:H. This result will be further investigated further via QE analysis and dark IV measurements.

Table 7.2.: Heterojunction solar cell characteristics for p-type and n-type cells passivated with either a-SiO$_x$:H(i) or a-Si:H(i). The thickness of the applied buffer layer (either a-SiO$_x$:H(i) or a-Si:H(i)) is given by d$_{buffer}$.

process	d$_{buffer}$ (nm)	Solar cell results J$_{sc}$ (mA/cm^2)	V$_{oc}$ (mV)	FF (%)	η (%)	R$_s$ (Ωcm^2)	R$_p$ (Ωcm^2)
n-type FZ-Si 1 Ωcm							
a-Si:H(i)	3.0	32.41	618	78.34	15.69	0.29	$\geq$ MΩ
a-SiO$_x$:H(i)	2.7	32.37	695	79.0	17.76	0.3	$\geq$ MΩ
p-type FZ-Si 0.5 Ωcm							
a-Si:H(i)	3.0	30.66	634	70.44	13.7	0.36	$\geq$ MΩ
a-SiO$_x$:H(i)	2.7	31.01	655	81.64	16.6	0.09	$\geq$ MΩ

Comparing those results to p-type based heterojunction cells, the V_{oc} for wafers passivated with a-SiO$_x$:H(i) and a-Si:H(i) is lower, namely 655 mV for a-SiO$_x$:H(i) and 634 mV for a-Si:H(i). This can be attributed to the lower effective lifetime ascertained for 0.5 Ωcm p-type wafers ($cf.$ section 6.6). The current for all samples is relatively low, due to the application of double-side polished wafer, which reveal a high optical reflectivity. Optical analysis in section 6.5.4 shows that the intrinsic a-SiO$_x$:H(i) films reveal a distinct decrease of blue light absorption compared to that of commonly used amorphous silicon (a-Si:H(i)). As a consequence, the fraction of light transmitted to the wafer which is available for solar conversion increases drastically. Figure 7.4 displays the quantum efficiency for heterojunction solar cells passivated with either a-SiO$_x$:H(i) or a-Si:H(i) for both n-type and p-type bulk material. The quantum efficiency is reduced due to recombination effects. The same mechanisms which affect the collection probability also affects the quantum efficiency, i.e., front surface passivation affects carriers generated near the surface. Since blue light is absorbed very close to the surface, high front surface recombination will affect the 'blue' portion of the quantum efficiency [2]. Figure 7.4 exhibits a distinct enhancement of the blue light response for the samples passivated with a-SiO$_x$:H compared to the standard a-Si:H(i). The enhancement of the blue light may be due to (i) the increased front surface passivation by the a-SiO$_x$:H or (ii) the increased fraction of light transmitted to the wafer, which now is available for solar conversion. Similarly, light in the long wavelength region is absorbed in the bulk of a solar cell and a low diffusion length would affect the collection probability from the solar cell bulk and reduce the quantum efficiency in the long wavelength portion of the spectrum. As seen in Fig. 7.4, the response in the long wavelength region for the n-type wafer passivated by a-SiO$_x$:H is favorably enhanced. This might be due to increased back surface passivation. Surprisingly, the response of the a-SiO$_x$:H passivated p-type wafer in the long wavelength region does not improve compared to standard a-Si:H(i).

Light- and dark-IV measurements of heterojunction solar cells for n-type and p-type material passivated with a-SiO$_x$:H (their structure is given in Fig. 7.3) has been carried out. The light-IV curves are outlined in Fig. 7.5a, and the dark-IV curves are presented in Fig. 7.5b. For comparison, heterojunction solar cells passivated with a-Si:H(i) with the same layer thickness as the a-SiO$_x$:H are superposed. It is found from the analysis of the dark characteristics of the n-type cell devices that the backward current density is reduced by 2 orders of magnitude when inserting a-SiO$_x$:H films for passivation. In the dark-IV curves shown in Fig. 7.5b, the diffusion current region shifts to a higher voltage for a-SiO$_x$:H passivated cells, and, at the same time, the reverse leakage current decreases with increase in V$_{oc}$. Comparing these results with the corresponding light-IV characteristics in Fig. 7.5a, the suppression of the backward current density improves the V$_{oc}$ up to 695 mV for n-type cells and hence the solar cell performance. It can be stated that this drastic increase is due to an excellent a-SiO$_x$:H surface passivation on the front- and backside of the cells.

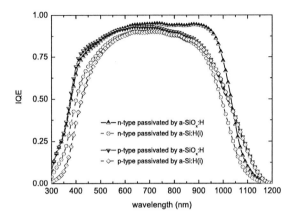

Figure 7.4.: IQE of heterojunction solar cells with detailed characteristics given in Table 7.2.

7.3.2. Influence of TCO on V_{oc}

Figure 7.6 shows the measured Suns-V_{oc} curve of a finalized n-type heterojunction solar cell passivated with a-SiO$_x$:H, as illustrated in the inset in Fig. 7.6. The 1-sun curve (light-IV) is superposed. An implied V_{oc} of 725 mV has been read from QSSPC measurements prior TCO deposition, as displayed in Fig. 7.6. The V_{oc} of the cell device is limited by surface recombination. The final TCO deposition does not enhance the recombination and introduces only small series resistance losses. But as seen from the Suns-V_{oc}, the ITO deposition is detrimental as it causes increased recombination and consequently diminishes the V_{oc}. The final comparison of the Suns-V_{oc} and the 1-sun IV curves indicates appreciable series resistance losses, *cf.* [158].

7.3.3. Impact of a-SiO$_x$:H layer thicknesses on cell performance

As already mentioned, the light absorption in standard a-Si:H(i) and the doped a-Si:H layers is rather strong, and the short-circuit current density in the a-Si:H/c-Si solar cells reduces rapidly with increasing a-Si:H layer thicknesses.

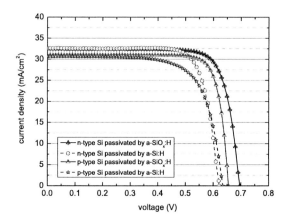

(a) Light IV characteristics.

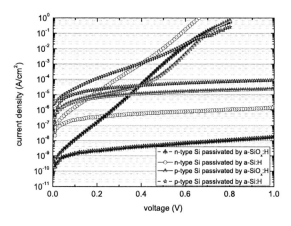

(b) Dark IV characteristics.

Figure 7.5.: IV characteristics of n-type and p-type heterojunction solar cells passivated with a-SiO$_x$:H (for cell structure, see Fig. 7.3). For comparison, the IV characteristics of n-type and p-type heterojunction solar cells passivated with standard a-Si:H(i) are superposed.

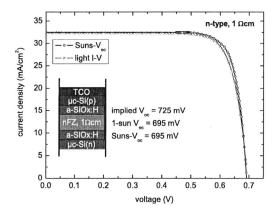

Figure 7.6.: Suns-V_{oc} curve of a finalized n-type heterojunction solar cell passivated with a-SiO$_x$:H. The 1-sun curve (light-IV) is superposed. The inset as illustrates the sample configuration. An implied V_{oc} of 725 mV has been read from QSSPC measurements prior TCO deposition, as displayed in Fig. 7.6. The V_{oc} of the cell device is limited by surface recombination. The final TCO deposition does not enhance the recombination and introduces only small series resistance losses.

To improve the short circuit density in a-Si:H/c-Si solar cells further, it is preferable to employ an a-Si:H based alloy that has a larger optical bandgap than standard a-Si:H. A promising alternative to a-Si:H is the application of a-SiO$_x$:H. The a-SiO$_x$:H layer introduced at the heterojunction, however, has been rather difficult to characterize due to the thin layer thickness on the nanoscale. Therefore, this experiment outlines the impact of the applied a-SiO$_x$:H thickness on the performance of heterojunction solar cells. The heterojunction cell design as shown in Fig. 7.3 is applied in this experiment.

7.3.3.1. Influence of a-SiO$_x$:H layer thickness symmetrically distributed at front- and backside

Figure 7.7 illustrates the impact of the applied a-SiO$_x$:H layer thickness as a function of V_{oc}, R_s, FF, J_{sc}, and η for both n- and p-type based heterojunction solar cell devices. The deposition parameters for both a-SiO$_x$:H and μc-Si:H layers are kept constant. From Fig. 7.7 can be seen that the film thickness is varied up to 5.5 nm. A further increase of the thickness lead to essential losses of the solar cell performance due to low electrical conductivity of the intrinsic amorphous silicon oxide layers. In Table 7.3 the cell results are summarized. The optimum film thickness is determined by a trade-off between passivation quality, current losses due to low conductivity of the a-SiO$_x$:H films resulting in high series resistances, light absorption and deposition homogeneity. It is found, that for p-type wafers, the optimum a-SiO$_x$:H film thickness in respect of cell performance appears to be around d$_{a-SiO:H}$ = 3.5 nm, whereas for n-type wafers the optimum thickness is found to be lower at d$_{a-SiO:H}$ = 2.5 nm.

Table 7.3.: Heterojunction solar cell characteristics for p-type and n-type cells passivated with a-SiO$_x$:H(i) of different thicknesses. The thickness of the applied buffer layer (a-SiO$_x$:H(i)) is given by d$_{a-SiO:H}$.

process	d$_{a-SiO:H}$ (nm)	J$_{sc}$ (mA/cm^2)	V$_{oc}$ (mV)	FF (%)	η (%)	R$_s$ (Ωcm^2)	R$_p$ (Ωcm^2)
			Solar cell results				
n-type FZ-Si 1 Ωcm							
a-SiO$_x$:H(i)	1.8	32.01	657	80.08	16.84	0.38	$\geq$ MΩ
a-SiO$_x$:H(i)	2.7	32.37	695	79	17.76	0.3	$\geq$ MΩ
a-SiO$_x$:H(i)	3.6	32.41	620	78.34	15.69	0.29	$\geq$ MΩ
a-SiO$_x$:H(i)	5.4	31.24	392	37.42	4.59	6.02	$\geq$ MΩ
p-type FZ-Si 1 Ωcm							
a-SiO$_x$:H(i)	1.8	30.44	651	71.27	14.13	0.17	$\geq$ MΩ
a-SiO$_x$:H(i)	2.7	30.42	665	75.91	15.34	0.39	$\geq$ MΩ
a-SiO$_x$:H(i)	3.6	31.03	655	81.64	16.6	0.09	$\geq$ MΩ
a-SiO$_x$:H(i)	5.4	29.36	630	68.63	12.67	2.5	$\geq$ MΩ

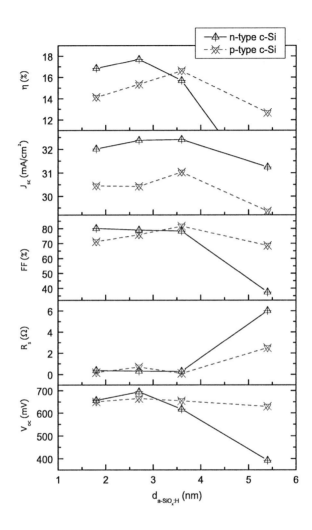

Figure 7.7.: Heterojunction solar cell performance depending on the applied a-SiO$_x$:H layer thickness. For *p*-type wafers, the optimum a-SiO$_x$:H film thickness in respect of cell performance appears to be around d$_{a-SiO:H}$ = 3.5 nm, whereas for *n*-type wafers the optimum thickness is found to be lower at d$_{a-SiO:H}$ = 2.5 nm.

7.3.3.2. Influence of a-SiO$_x$:H layer thickness asymmetrically distributed at front- and backside

The a-SiO$_x$:H layer thickness is a crucial parameter for the proper performance of the heterojunction solar cell devices. Although the incorporation of the a-SiO$_x$:H layer at the a-Si:H/c-Si hetero-interface has been confirmed to improve the solar cell performance, the role of the thin a-SiO$_x$:H layer in the a-Si:H/c-Si solar cells still remains ambiguous. Such a characterization is of significant importance to clarify the role of the a-SiO$_x$:H layer in the heterojunction cell devices. In particular, the characteristics of the heterojunction solar cells will be tested in the next experiment with a variation of the thickness of the a-SiO$_x$:H layer. Therefore, an asymmetric distribution of a-SiO$_x$:H on front- and backside is applied. The front side will be covered by a-SiO$_x$:H layer of constant thickness (optimum a-SiO$_x$:H film thickness in respect of cell performance appears to be $d_{a-SiO:H}$ = 3.5 nm to 5 nm for p-type, and $d_{a-SiO:H}$ = 2.5 nm for n-type based cells), whereas the backside will be covered by a-SiO$_x$:H of different thicknesses ($d_{a-SiO:H}$ = 1.5 nm, 2.1 nm, 3.3 nm, and 3.9 nm). The heterojunction cell design used is shown in Fig. 7.3. All other parameters are constant.

Figure 7.8 shows the heterojunction solar cell performance as a function of the applied a-SiO$_x$:H layer thickness on the backside. Accordingly, the optimum a-SiO$_x$:H film thickness at the backside in respect of cell performance appears to be around $d_{a-SiO:H}$ = 3.5 nm for p-type cells, and $d_{a-SiO:H}$ = 2.7 nm for n-type cells. Table 7.4 summarizes the results.

The results imply that the carrier transport and recombination in the heterojunction solar cells vary significantly with the a-SiO$_x$:H(i) layer thickness. For completeness, the corresponding light- and dark-IV curves are outlined in Fig. 7.9 for n-type cells, and in Fig. 7.10 for p-type cells. It is known that the dark forward current, an hence the V$_{oc}$ is affected by the presence of localized recombination sources within a diffusion length of the junction [2]. Thus, as seen in Fig. 7.9 and Fig. 7.10, the impact of surface recombination, which is reduced by passivating the surfaces, is a function of passivation layer thickness. Hence, the thickness of a-SiO$_x$:H needs to be optimized separately at the front- and backside.

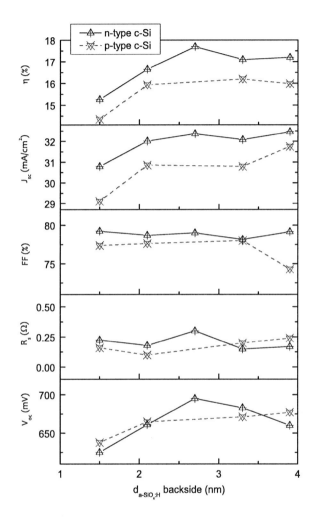

Figure 7.8.: Heterojunction solar cell performance as a function of the applied a-SiO$_x$:H layer thickness on the backside. The front side is covered by a constant a-SiO$_x$:H layer of d$_{a-SiO:H}$ = 2.7 nm for n-type and d$_{a-SiO:H}$ = 3.6 nm for p-type wafers. The optimum a-SiO$_x$:H film thickness at the backside in respect of cell performance appears to be around d$_{a-SiO:H}$ = 3.5 nm for p-type cells, and d$_{a-SiO:H}$ = 2.7 nm for n-type cells.

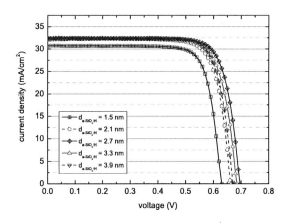

(a) Light IV characteristics.

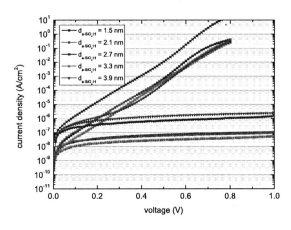

(b) Dark IV characteristics.

Figure 7.9.: IV characteristics of n-type heterojunction solar cells passivated with a-SiO$_x$:H of different thicknesses on the backside. (for cell structure, see Fig. 7.3). For comparison, the IV characteristics of n-type and p-type heterojunction solar cells passivated with standard a-Si:H(i) are superposed.

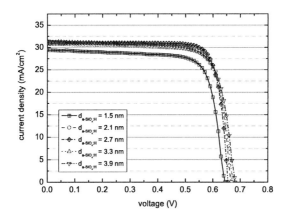

(a) Light IV characteristics.

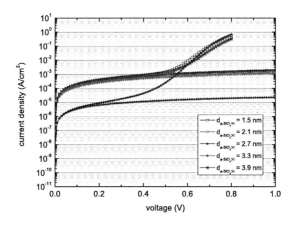

(b) Dark IV characteristics.

Figure 7.10.: IV characteristics of p-type heterojunction solar cells passivated with a-SiO$_x$:H of different thicknesses on the backside (for cell structure, see Fig. 7.3). For comparison, the IV characteristics of n-type and p-type heterojunction solar cells passivated with standard a-Si:H(i) are superposed.

Table 7.4.: Heterojunction solar cell characteristics for *p*-type and *n*-type cells. The front side is passivated by an a-SiO$_x$:H layer of constant thickness (optimum a-SiO$_x$:H film thickness in respect of cell performance appears to be $d_{\text{a-SiO:H}}$ = 3.5 nm to 5 nm for *p*-type, and $d_{\text{a-SiO:H}}$ = 2.5 nm for *n*-type based cells), whereas the backside is covered by a-SiO$_x$:H of different thicknesses ($d_{\text{a-SiO:H}}$ = 1.5 nm, 2.1 nm, 3.3 nm, and 3.9 nm). The thickness of the applied buffer layer on the backside (a-SiO$_x$:H) is given by $d_{\text{a-SiO:H}}$.

		Solar cell results					
process	$d_{\text{a-SiO:H}}$ (nm)	J_{sc} (mA/cm^2)	V_{oc} (mV)	FF (%)	η (%)	R_s (Ωcm^2)	R_p (Ωcm^2)
_n_-type FZ-Si 1 Ωcm							
a-SiO$_x$:H(i)	1.5	30.79	625	79.23	15.26	0.22	$\geq$ MΩ
a-SiO$_x$:H(i)	2.1	32.02	661	78.68	16.65	0.18	$\geq$ MΩ
a-SiO$_x$:H(i)	2.7	32.37	695	79	17.7	0.3	$\geq$ MΩ
a-SiO$_x$:H(i)	3.3	31.08	683	78.14	17.1	0.15	$\geq$ MΩ
a-SiO$_x$:H(i)	3.9	31.46	660	79.15	17.2	0.17	$\geq$ MΩ
_p_-type FZ-Si 1 Ωcm							
a-SiO$_x$:H(i)	1.5	29.11	638	77.38	14.36	0.16	$\geq$ kΩ
a-SiO$_x$:H(i)	2.1	30.86	665	77.63	15.93	0.1	$\geq$ kΩ
a-SiO$_x$:H(i)	3.3	30.8	671	78.01	16.2	0.2	$\geq$ kΩ
a-SiO$_x$:H(i)	3.9	31.76	677	74.3	15.98	0.24	$\geq$ kΩ

7.3.3.3. Influence of unpassivated sidewalls

Transport mechanisms that might be relevant to the heterojunction solar cell devices are multi-tunneling capture emission, tunneling at unpassivated sidewalls, and the effect of the series resistance. Since the sidewalls of the heterostructures are left unpassivated, some leakage current is expected to appear in the dark IV measurements. Since the tunneling at the sidewall defects increases with increasing electric field, this transport mechanism may be the dominant transport mechanism in a reverse-bias regime. However, this effect is not expected to affect the performance of the solar cells because solar cells operate at a medium forward-bias regime, where the electric field in the junction is considerably smaller than the reverse-bias regime [90].

7.3.4. Impact of post-annealing of emitter and BSF on heterojunction solar cell performance

This section focuses on the impact of post-annealing on heterojunction solar cell performance. This is related to the fact that, as shown in chapter 5, the p^+ and n^+ μc-Si:H layer benefits from post-annealing. Subsequent low-temperature post deposition annealing for film defect reduction is carried out as shown in the flowchart in Fig. 7.2. In case the intrinsic a-SiO$_x$:H buffer layer is covered with a doped μc-Si:H overlayer, care has to be taken however when applying such annealing step. This is due to the fact that the presence of such doped layer may lower the energy required for Fermi-level dependent Si-H bond rupture in the underlying intrinsic buffer layer, resulting in degraded passivation [142]. Consequently, optimal passivation results of stacked doped μc-Si:H/a-SiO$_x$:H/c-Si structures are suggested to be obtainable by deposition of the intrinsic a-SiO$_x$:H buffer layer, which is followed by post-deposition annealing to lower the defect-density of the films. Subsequently, the doped μc-Si:H layers are deposited, which is followed by a second post-deposition annealing to lower the defect-density of the films, as proven already in sections 5.4.2 and 5.3.2. Figure 7.11 shows the results of heterojunction solar cells performance depending on the post-annealing temperature after the deposition of emitter and BSF. An optimum post-annealing temperature of $180\,^\circ$C can be determined.

7.4. Application of a-SiO$_x$:H and μc-Si:H films to heterojunction solar cells using textured float-zone c-Si

In combination with the anti-reflection coating, ITO (*cf.* appendix A), texturing reduces the reflection losses drastically. Light entering the substrate at the textured surface is then tilted with respect to the cell's local surface normal. This means that the probability light enters the cell, which would be normally reelected at the front side or backside, gets another chance to hit the cells surface or for internal reflection. Also, the photogeneration takes place closer to the collection junction, enhancing the collection efficiency of medium- to long wavelengths, which is beneficial for low-diffusion length a-Si:H material [1].

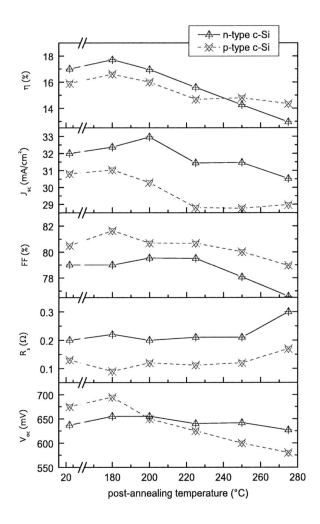

Figure 7.11.: Heterojunction solar cell performance depending on the post-annealing temperature after deposition of emitter and BSF. For both *p*-type and *n*-type based solar cells, the optimum post-annealing temperature after deposition of emitter and BSF is found to be around 180 °C.

The same effect would be gained with an increase of the absorption coefficient. The main drawback of textured surfaces is that those present higher surface recombination velocities. In the following, this effect will be counteracted by applying a-SiO$_x$:H as a buffer layer.

7.4.1. Impact of emitter layer thicknesses on cell performance

The applied heterojunction design is illustrated in the cross sectional view displayed in Fig. 7.3. Identically deposition parameters as for polished c-Si heterojunction solar cells are used. Table 7.5 summarizes the results for the photovoltaic output parameters of textured heterojunction solar cells passivated with a-SiO$_x$:H(i) for n-type bulk material and various emitter thicknesses, d$_{emitter}$. It is found that a optimum μc-Si:H(p) emitter thickness around d$_{emitter}$ = 20 nm leads to an efficiency exceeding 19 % (confirmed at ISE CalLab). To investigate this gain in efficiency further, the inpact of a-SiO$_x$:H layer thicknesses on the cell performance is dicussed in the next section.

Table 7.5.: Heterojunction solar cell characteristics for textured n-type cells passivated with a-SiO$_x$:H for various μc-Si:H emitter layer thicknesses. The thickness of the applied emitter layer is given by d$_{emitter}$. The thickness of the applied passivation layer is kept constant with d$_{a\text{-}SiO:H}$ = 2.7 nm.

		Solar cell results					
process	d$_{emitter}$ (nm)	J$_{sc}$ (mA/cm^2)	V$_{oc}$ (mV)	FF (%)	η (%)	R$_s$ (Ωcm^2)	R$_p$ (Ωcm^2)
n-type FZ-Si 1 Ωcm							
	15.0	36.90	668	77.50	19.01	0.29	$\geq$ MΩ
	20.0	37.45	660	77.61	19.18	0.20	$\geq$ MΩ
	22.5	37.20	660	72.51	17.80	0.38	$\geq$ MΩ
	25.0	36.04	667	76.56	18.21	0.34	$\geq$ MΩ

7.4.2. Impact of a-SiO$_x$:H layer thicknesses on cell performance

Figure 7.3 illustrates a cross sectional view of the heterojunction design applied, which consists of basically the same deposition parameters as are used

for polished c-Si heterojunction solar cells. Table 7.6 summarizes the results
for photovoltaic output parameters of textured heterojunction solar cells passivated with a-SiO$_x$:H(i) for n-type bulk material. Comparing those results
to the results achieved on polished substrates, the expected gain of J$_{sc}$ from
32.5 mA/cm^2 up to ~37. mA/cm^2 can be observed. The transfer of the same
process parameters used for optimized heterojunction cells with polished
wafers to heterojunction cells with textured wafers leads to a decrease in V$_{oc}$.
The V$_{oc}$ drops from 695 mV obtained on polished substrates to 660 mV for
textured substrates. This result is not surprising as the required layer thickness of a-SiO$_x$:H will be not obtained by using the same deposition time, as
found already in [159]. Adjusting the thickness of the a-SiO$_x$:H layers results
in an optimized film thickness of d$_{a-SiO:H}$ ≥ 2.7 nm covering the randomly
distributed pyramids. The V$_{oc}$ increases from 660 mV to 675 mV.

Table 7.6.: Heterojunction solar cell characteristics for textured n-type cells passivated
with a-SiO$_x$:H. The thickness of the applied buffer layer is given by d$_{a-SiO:H}$.

process	d$_{a-SiO:H}$ (nm)	J$_{sc}$ (mA/cm^2)	V$_{oc}$ (mV)	FF (%)	η (%)	R$_s$ (Ωcm^2)	R$_p$ (Ωcm^2)
			Solar cell results				
n-type FZ-Si **1 Ωcm**							
a-SiO$_x$:H(i)	2.3	37.25	659	76.79	18.9[a]	0.20	≥ MΩ
a-SiO$_x$:H(i)	2.7	37.03	675	77.3	19.3[a]	0.20	≥ MΩ

[a] confirmed at ISE CalLab.

Light- and dark-IV measurements of textured n-type heterojunction solar
cells passivated with a-SiO$_x$:H have been carried out. The light-IV curves are
outlined in Fig. 7.12a, whereas the dark-IV curves are presented in Fig. 7.12b.
For comparison, heterojunction solar cells using polished substrates are superposed. It is found from the analysis of the dark characteristics of the n-type
cell devices that the backward current density of the textured cell is slightly
higher. This might be due to inadequately passivated pyramids on the substrate. The quantum efficiency of those heterojunction cells are shown in

Fig. 7.13. For comparison, the IV characteristics of polished n-type hetero-junction solar cells passivated with either a-SiO$_x$:H or a-Si:H are superposed.

Optical analyses (*cf*. section 6.5.4) showed that the intrinsic a-SiO$_x$:H films reveal a distinct decrease of blue light absorption compared to that of commonly used a-Si:H. As a consequence, the fraction of light transmitted to the wafer which is available for solar conversion increases. The quantum efficiencies, displayed in Fig. 7.13 exhibit a distinct increase in the short wavelength region for a-SiO$_x$:H passivated wafers, both the polished and textured substrate, compared to standard a-Si:H. This yield can be attributed to (i) excellent surface passivation at the front side and (ii) the increased fraction of light transmitted to the wafer, which now is available for solar conversion.

The spectral response in the long wavelength (*cf*. Fig. 7.13) region for both the n-type polished and texturized wafer passivated by a-SiO$_x$:H is favorably enhanced. This might be due to increased back surface passivation. Comparing the long wavelength region response closer, a slight decrease of quantum efficiency of the texturized wafer based cell can be attributed to the inadequately passivated pyramids on the substrate. The loss in V_{oc} (675 mV for texturized substrates compared to 695 mV for polished substrates) and FF confirms that result. For future work, the a-SiO$_x$:H film thickness needs to be further optimized for adequately and homogeneously pyramid surface coverage.

7.5. Chapter summary

The present chapter reports on the results of a-SiO$_x$:H passivation incorporated into heterojunction solar cells for high quality surface passivation. Intrinsic a-SiO$_x$:H(i) films are formed in order to prove their applicability for surface passivating buffer layers sandwiched between the crystalline silicon (c-Si) and the doped μc-Si:H layer used for the formation of the emitter and the back-surface-field in heterojunction cells.

By incorporating a-SiO$_x$:H(i) to the heterojunction structure a drastic increase of the open circuit voltage (up to 655 mV for p-type substrates and 695 mV for n-type substrates) is found, and accordingly, a higher conver-

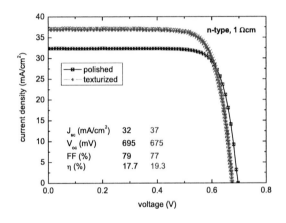

(a) Light IV characteristics.

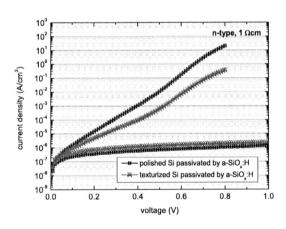

(b) Dark IV characteristics.

Figure 7.12.: IV characteristics of textured *n*-type heterojunction solar cells passivated with a-SiO$_x$:H (for cell structure, see Fig. 7.3). For comparison, the IV characteristics of polished *n*-type heterojunction solar cells are superposed.

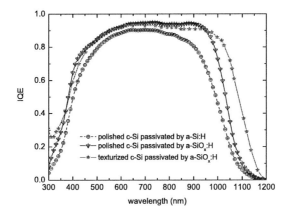

Figure 7.13.: Quantum efficiency of textured n-type heterojunction solar cells passivated with a-SiO$_x$:H (for cell structure, see Fig. 7.3). For comparison, the IV characteristics of polished n-type heterojunction solar cells passivated with either a-SiO$_x$:H or a-Si:H are superposed.

sion efficiency than obtained with standard a-Si:H(i). It has been shown that a-SiO$_x$:H surface passivation scheme is an adequate alternative for standard a-Si:H(i). One remarkable feature of the heterojunction solar cells investigated in this chapter is the extremely thin a-SiO$_x$:H layers employed in an a-SiO$_x$:H/c-Si structure. Thus, in spite of the importance of the a-SiO$_x$:H layer, almost no studies have focused on the structure and properties of the thin a-SiO$_x$:H layer incorporated in actual solar cell devices. Such characterization is of significant importance to clarify the role of the a-SiO$_x$:H layer in the heterojunction cell devices. It is found that the surface passivation quality depends on the applied layer thicknesses, in particular this becomes true for heterojunction solar cells. The optimum film thickness thereof is determined by a trade-off between passivation quality, current losses due to low conductivity of the a-SiO$_x$:H films resulting in high series resistances, light absorption and deposition homogeneity. It is found, that for p-type wafers, the optimum a-SiO$_x$:H film thickness in respect of cell performance appears to be around $d_{a-SiO:H} = 3.5$ nm, whereas for n-type wafers the optimum thickness is found to be around $d_{a-SiO:H} = 2.5$ nm. The responding IQE exhibited a distinct decrease in the short wavelength region for a-SiO$_x$:H passivated wafer of both doping types compared to standard a-Si:H(i). These high open-circuit voltages can be consistently ascribed to the adequate surface passivation by a-SiO$_x$:H preventing surface recombination at the hetero-interface and to the decrease of the optical absorption in the blue light region due to an enhanced optical bandgap of 1.95 eV.

Heterojunction solar cells using textured substrates exhibited an expected J_{sc} gain, when transferring the optimized process parameters of cells using polished substrates to cells using textured substrates. The thickness of the a-SiO$_x$:H layers needs to be adjusted to result in a film thickness of $d_{a-SiO:H} \geq 2.7$ nm covering the randomly distributed pyramids.

CHAPTER 8.

CONCLUDING REMARKS AND RECOMMENDATIONS FOR FUTURE RESEARCH

8.1. General conclusions

The work in this thesis is bifided, though it aims one purpose: the optimization of heterojunction solar cell devices. The key role in obtaining high efficient heterojunction solar cells are mainly: (i) the PECV deposition of very low defect layers, and (ii) the sufficient surface passivation of all interfaces occurring in the heterostructure cell device. In general, this thesis concerns about various types of PECV deposited alloys of silicon, hydrogen, carbon, oxygen and doping source gases.

Hydrogenated amorphous silicon (a-Si$_x$:H$_y$) has been accepted as a suitable heterojunction material for a-Si:H/c-Si solar cells, and conversion efficiencies exceeding 21 % has been reported recently [22]. Nevertheless, photon absorption in the emitter of a heterojunction solar cells leads to a considerable current loss with increasing a-Si:H layer thickness due to the high recombination in this layer [123]. To improve J_{sc} in a-Si:H/c-Si solar cells, it is preferable to employ an a-Si:H-based alloy that has larger optical bandgap than a-Si:H to suppress light absorption in the window layer or to reduce the recombination. Therefore, one part of this work concentrates on the development and investigation of PECV deposited wide-gap films serving as high transparent alternative to standard amorphous silicon (a-Si:H) used in heterojunction cell devices as emitter and back-surface-field; in particular, PECV deposited amorphous silicon carbon alloys (a-Si$_x$C$_{1-x}$:H$_y$), and micro-crystalline silicon (μc-Si:H) layers have been compared to standard a-Si:H.

The second part of this thesis concentrates on the development and investigation of a high quality crystalline silicon (c-Si) surface passivation to enhance

the hetero-interface between the a-Si:H/c-Si hetero device. In the research field
of crystalline silicon solar cells, electronic surface passivation has been recognized as a crucial step to achieve high conversion efficiencies. High bulk and
surface recombination rates are known to limit the open circuit voltage and
to reduce the fill-factor of photovoltaic devices [26, 33]. The suppression of
surface recombination by applying any kind of surface passivation scheme is
thereby one of the basic prerequisites to obtain high efficiency solar cells. This
becomes particularly true for heterojunction solar cells: featuring an abrupt
discontinuity of the crystal network at the crystalline silicon (c-Si) surface to
the amorphous emitter (a-Si:H) results usually in a large density of defects in
the bandgap due to a high density of dangling bonds [60]. These defects at the
hetero-interface encounter often detrimental effects on the solar cell performance, *cf*. [140]. Although the electrical field can reduce the recombination
near the hetero-interface, the junction properties are still governed by the interface state density. Therefore, in order to obtain high efficiency solar cells
it is essential to reduce the interface state density [9]. In particular, this has
been carried out by PECV deposited amorphous silicon sub-oxides (a-SiO$_x$:H).
While the investigations on wide-gap PECV deposited a-Si$_x$C$_{1-x}$:H$_y$ and μc-Si:H
films has focused on theirs application on heterojunction solar cell devices,
the a-SiO$_x$:H surface passivation scheme is intended for broader applications
in general silicon solar cell devices.

8.1.1. Wide-gap a-Si$_x$C$_{1-x}$:H$_y$ layers for use as emitter

Hydrogenated amorphous alloys of silicon and carbon (a-Si$_x$C$_{1-x}$:H$_y$) are an
interesting variant to standard a-Si:H used in heterojunction solar cell devices.
The addition of carbon adds extra freedom in order to control the properties
of the material, as increasing concentrations of carbon in the alloy widen the
optical bandgap in order to potentially increase the light efficiency of solar
cells made with amorphous silicon carbide layers. It is found that by proper
valency control amorphous silicon carbide can be utilized to widen the band
gap of the window layer of the heterojunction cell. The alloys are fabricated by
decomposition of silane (SiH$_4$), phosphine (PH$_3$), methane (CH$_4$) and hydrogen
(H$_2$), using a PECVD setup. Particularly, the investigation focused on the in-

corporation of hydrogen and carbon within the resulting a-Si$_x$C$_{1-x}$:H$_y$(n) and a-Si$_x$:H$_y$(n) films, which later form the emitter. The corresponding local vibrational modes of Si-H$_x$, C-H and corresponding network have been analyzed by μ-Raman spectroscopy. The addition of carbon degrades the photoelectronic properties in the emitter layer. This deterioration can be minimized by H dilution. The resulting optical band gap E$_G$ as well as the thickness of the emitter are determined by spectroscopic ellipsometry. It is confirmed that the band gap E$_G$ can be tailored by using an appropriate gas mixture during the decomposition. Furthermore the I-V characteristics of the prepared heterojunction solar cells are analyzed. A trade-off between electrical defects density and optical losses induced an improvement of the I-V characteristics with increasing carbon and hydrogen concentration in the feed stock during deposition. However, there are problems such as low conversion efficiency due to the sp^2/sp^3 bonding structure of C-C [129], as well as deterioration of the photo-conductivity, because of carbon induced electronic defects [129]. The density of defect states should be minimized by hydrogen dilution [160].

8.1.1.1. Recommendations for future research

As for amorphous silicon thin film cells, i.e., the use of a-SiC:H deposited from silane and methane source gases is an effective technique to achieve higher conversion efficiencies in amorphous silicon solar cells by PECVD. Of course, these films show a high defect density and a fast oxidation in air. One possible approach to improve this deterioration would a change of the gaseous precursor for carbon. In *Nonomura et al.* [161] a saturated hydrocarbon, butane (C$_4$H$_{10}$), is used as the carbon source gas in order to find a solution to the above problem. The bonding energies of C-H and C-C in C$_4$H$_{10}$ are the smallest among the saturated hydrocarbon gases. These low binding energies increase the dissociation of the carbon source gas and the carbon-based radicals so created improve the carbon incorporation rate into the films during the deposition.

8.1.2. Wide-gap, high-conductive μc-Si:H layers for use as emitter and BSF

According to the film properties discussed, the p^+ and n^+ μc-Si:H films are likely to be suitable for use as emitter and BSF in a heterojunction solar cell device. They indicated high transparency to suppress absorption, and high conductivity when annealed at the optimum temperature. For p^+ μc-Si:H layers, the dark conductivity increases with increasing TMB concentration till a maximum at χ_{TMB} = 2.62 %, corresponding to approximatly 30000 ppm. An optimum heater temperature appropriate for high conductivity is found at T_{heat} = 280 °C for the PECVD setup used, corresponding to a deposition temperature of T_{dep} = 140 °C. It is found that the fraction of microcrystallinity can be varied as a function of hydrogen dilution. Low χ_H rates result in a larger fraction of amorphous structures inside the deposited film with Raman peaks around 480 cm^{-1}, whereas high H_2 dilution (χ_H > 98 %) results in a more micro-crystalline character with peaks around 500 cm^{-1}. The μc-Si:H(p) layers indicated at optimal PECV deposition parameters high conductivities of σ_{dark} = 10 S/cm (corresponding to 0.1 Ωcm). The n^+ μc-Si:H films indicated significant higher dark conductivity values (with optimum process parameters approximately σ_{dark} = 100 S/cm) compared to the σ_{dark} values of p^+ μc-Si:H films. The doping concentration for the highest σ_{dark} is found to be approximately χ_{PH_3} = 1.5 %. It can be concluded that the fraction of micro-crystallinity can be varied the plasma frequency, and is also influenced by the deposition temperature and hydrogen dilution. It is found that the doping efficiency and carrier mobility are optimal at VHF plasma frequency of 110 MHz, and with high hydrogen dilution of χ_H = 98 %. The dark conductivity σ_{dark} and hence the carrier mobility are improved by means of thermal annealing for both n^+ and p^+ μc-Si:H films. The μc-Si:H(p) films annealed with temperatures in the range of 200 °C < T_{ann} < 375 °C are appropriate to obtain the high dark conductivity values (with optimum process parameters results in approximately σ_{dark} = 25 S/cm, corresponding to 0.04 Ωcm). By means of post-annealing of the μc-Si:H films, an activation of the boron atoms occurs, which was formerly bound in form of B-H-Si. The annealing step might release the hydrogen from this B-H-Si bound, leaving activated B-Si bounds. For μc-Si:H(n) films an annealing temperature in the range between $180\,°C < T_{ann} < 300\,°C$ is found to be an optimum temperature to improve dark conductivity up to $\sigma_{dark} = 130\,S/cm$ (corresponding to $0.0077\,\Omega cm$).

8.1.3. Surface recombination of c-Si passivated with PECVD a-SiO$_x$:H

A novel passivation scheme, featuring PECV deposited hydrogenated amorphous silicon sub-oxide films (a-SiO$_x$:H), has been investigated extensively. The a-SiO$_x$:H films are deposited by decomposition of silane (SiH_4), carbon dioxide (CO_2) and hydrogen (H_2) as source gases. The plasma deposition parameters of a-SiO$_x$:H films were optimized in terms of effective lifetime, whereas the oxygen content and the resulting optical band gap E_{04} of the a-SiO$_x$:H films were controlled by varying the CO_2 partial pressure $[CO_2]/([CO_2]+[SiH_4])$. An optimum substrate deposition temperature and plasma frequency of $T_{dep} = 155\,°C$ and 70 MHz, respectively, have been revealed. Also, an optimized gas flow ratio of $\chi_{ox} = 1{:}5$ has been employed. Additional H_2 reduces the oxygen content to 10 at.% in the precursor flow during deposition.

The impact of post-annealing of low temperature annealing at 250 °C of those films showed a beneficial effect in form of a drastic increase of the effective lifetime. This improvement of the passivation quality by low temperature annealing for the a-SiO$_x$:H likely originates from defect reduction of the film close to the interface. Thermal annealing at higher temperature (above 450 °C) leads to a decreasing lifetime and to an effusion of hydrogen. Raman spectra reveal the existence of Si-(OH)$_x$ and Si-O-Si bonds after thermal annealing of the layers, leading to a higher effective lifetime, as it reduces the defect absorption of the sub oxides.

The surface passivation of a-SiO$_x$:H within both *n*-type and *p*-type silicon has been studied as a function of injection level. Record high effective lifetime values of 4.7 ms on 1 Ωcm *n*-type FZ wafers and 14.2 ms on 130 Ωcm p-type FZ wafers prove the applicability for a surface passivation of silicon wafers to any kind of silicon based solar cells. These values appears to be the highest ever reported on a high doped crystalline wafer of 1 Ωcm resistivity. Samples prepared in this way feature a high quality passivation yielding effective lifetime values exceeding those of record SiO_2 and SiN_x passivation schemes. Main key factors to obtain such high quality films comprehend: (i) the chamber contamination, which depend on anteceded processes with various precursor gases, (ii) dummy process to cover the reactor walls with a-SiO_x:H, (iii) the deposition temperature and the pre-heating time in the chamber, and (iv) the electrode distance. The use of a-SiO_x:H may be a promising alternative for any passivation scheme existing so far. The only drawback found is the upper limit temperature stability for films prepared and tested in this work should not exceed a critical temperature of 500 °C. Further advantages of the a-SiO_x:H passivation scheme are that the fabricated a-SiO_x:H layers are grown by simple PECV deposition at low temperatures, and they withstand hydrofluoric acid (tested with a 5 % diluted HF). It has to be mentioned that the process parameters have to be carefully adjusted to obtain such high quality a-SiO_x:H films. In addition, a high-transparent window layer up to 2.4 eV, depending on the fraction of oxygen in the precursor gas, is obtained. Therefore, the a-SiO_x:H films exhibit low blue light absorption compared to a-Si:H(i), enabling a thicker passivation layer than standard a-Si:H(i) and therefore an increasing passivation quality.

8.1.3.1. Recommendations for future research

It has to be mentioned that the process parameters have to be carefully adjusted to obtain such high quality a-SiO_x:H films. The maximum surface area for c-Si passivation with a-SiO_x:H in this work has been limited to 10 by 10 cm due to the PECVD chamber size. For industrial processes the PECV deposition of a-SiO_x:H on large area is still a challenge, especially regarding to the optimized plasma frequency of 70 MHz.

8.1.4. Heterojunction solar cell structures passivated with a-SiO_x:H

The applicability as a high quality passivation scheme for use in heterojunction solar cells replacing the standard a-Si:H(i) buffer layer has been investigated. It has been shown that a-SiO$_x$:H surface passivation scheme is an adequate alternative for standard a-Si:H(i). It is found that the surface passivation quality depends on the applied layer thicknesses, in particular this becomes true for the application of those films in heterojunction solar cells. Intrinsic a-SiO$_x$:H(i) films are formed in order to prove their applicability for surface passivating buffer layers sandwiched between the crystalline silicon (c-Si) and the doped μc-Si:H layer used for the formation of the emitter and the back-surface-field in heterojunction cells. One remarkable feature of the heterojunction solar cells investigated is the extremely thin a-SiO$_x$:H layers employed in an a-SiO$_x$:H/c-Si structure. Thus, in spite of the importance of the a-SiO$_x$:H layer, almost no studies have focused on the structure and properties of the thin a-SiO$_x$:H layer incorporated in actual solar cell devices. Such characterization is of significant importance to clarify the role of the a-SiO$_x$:H layer in the heterojunction cell devices. It is found that the surface passivation quality depends on the applied layer thicknesses, in particular this becomes true for heterojunction solar cells. The optimum film thickness thereof is determined by a trade-off between passivation quality, current losses due to low conductivity of the a-SiO$_x$:H films resulting in high series resistances, light absorption and deposition homogeneity. It is found, that for p-type wafers, the optimum a-SiO$_x$:H film thickness in respect of cell performance appears to be around d$_{a-SiO:H}$ = 3.5 nm, whereas for n-type wafers the optimum thickness is found to be around d$_{a-SiO:H}$ = 2.5 nm. The responding IQE exhibited a distinct decrease in the short wavelength region for a-SiO$_x$:H passivated wafer of both doping types compared to standard a-Si:H(i). These high open-circuit voltages can be consistently ascribed to the adequate surface passivation by a-SiO$_x$:H preventing surface recombination at the hetero-interface and to the decrease of the optical absorption in the blue light region due to an enhanced optical bandgap of 1.95 eV.

By incorporating a-SiO$_x$:H(i) to the heterojunction structure a drastic increase of the open circuit voltage (up to 655 mV for p-type substrates and 695 mV for n-type substrates) is found, and accordingly, a higher conversion efficiency than obtained with standard a-Si:H(i). Heterojunction solar cells using textured substrates exhibited an expected J$_{sc}$ gain, when transferring the optimized process parameters of cells using polished substrates to cells using textured substrates. The thickness of the a-SiO$_x$:H layers needs to be adjusted to result in a film thickness of d$_{a-SiO:H}$ $\geq$ 2.7 nm covering the randomly distributed pyramids. Efficiencies exceeding 19 % have been reached.

8.1.4.1. Recommendations for future research

Key factors for high efficiency heterojunction solar cell devices resulting from this work:

- A high optical bandgap of the investigated a-SiC:H, a-SiO$_x$:H, and μc-Si:H layers results in low activation energy and low optical absorption (suppresses light absorption), therefore thicker intrinsic passivation and doped layers possible.

- The use of wide-gap a-SiO$_x$:H allows a high quality surface passivation.

- The use of high conductive wide-gap μc-Si:H layers used as emitter and BSF are mainly responsible for current gain.

- An interface between c-Si and doped emitter should be abrupt; namely amorphous.

APPENDIX A.

ANCILLARY NOTES REGARDING SAMPLE PREPARATION

Unless stated different elsewhere, all solar cells prepared and published in this thesis are subject to constant process steps, which include: (i) cleaning, (ii) transparent conductive layer, and (iii) metal contact fabrication process. Those processes are briefly described in terms of completion as follows.

A.1. Substrate pre-treatment

A.1.1. Cleaning

In order to create an excellent hetero-interface, it is essential to use well cleaned c-Si wafer surface and remove the native oxide prior loading to the PECVD system. In general, various processing techniques for cleaning can be applied prior plasma deposition. In this work the commonly used RCA cleaning as described in [162] is applied followed by a dip in hydrofluoric acid (HF) in order to remove any native oxide from the wafer surface prior PECV deposition. Other works have shown (i.e. [163]) that a RCA cleaning process could be avoided, as the commercial wafers are already well pre-cleaned and exhibit only a native oxide of a few Angstroms; but to ensure comparability of different sample series, a RCA cleaning is applied as a precaution to all samples in this work (unless stated different).

The RCA cleaning method is based on two solutions: standard-clean-1 (SC-1), an ammonium hydroxide/hydrogen peroxide/DI water mixture, and standard-clean-2 (SC-2), a hydrochloric acid/hydrogen peroxide/DI water mixture. Detailed process parameters for the cleaning procedure applied to the samples fabricated in this work are found in Table A.1.

Organic compounds are removed by sulfuric peroxide; ammonium hydrox-

Table A.1.: Basic conditions of RCA cleaning.

cleaning process	chemical composition	chemical ratios	temperature (°C)	process time
HF-dip	HF : H_2O	5 %	RT	10 s
SC-1	NH_4OH : H_2O_2 : H_2O	1 : 1 : 5	80	15 min
SC-2	HCl : H_2O_2 : H_2O	1 : 1 : 6	85	15 min
HF-dip	HF : H_2O	5 %	RT	30 s

ide removes particles; metallic impurities are removed by hydrochloric hydroxide; diluted HF removes the native oxide films; and rinsing is done with de-ionized (DI) water. The mechanisms of hydrogenation of Si surfaces in fluoride containing solutions (diluted HF) leads also to a H-termination of the c-Si surface.

After rinsing the wafer with DI-water, the sample is immediately transferred into the plasma deposition chamber for deposition of a-Si:H or its counterparts by means of rf-PECVD. The time period between the last HF-dip and the transfer into the PECVD chamber averages less than 5 minutes. It is assumed that during this procedure the H-termination still presents and re-growth of native oxides is reduced to a minimum, *cf*. [164].

A.1.2. Texturing

Chemical solutions can etch anisotropically amorphous Si crystal exposing {111} planes. On [100]-oriented wafers, randomly distributed square-base pyramids are formed. Light entering the substrate at the textured surface is then tilted with respect to the cell normal [1]. The samples textured in this thesis (if self-fabricated) are subject to a tetramethylammonium hydroxide (TMAH) etching solution. It consists of 2 % diluted in DI-water with 8 % IPA. The size of the pyramids is adjusted to a few microns by controlling the etching time and temperature (usually around 80 °C).

A.2. Transparent conductive oxide

The transparent conductive oxide (TCO) fulfills two purposes: (i) it serves as an anti-reflection coating (ARC) and (ii) it increases the lateral conductivity. Reflection is at a minimum when the layer thickness is an odd multiple of $\lambda_0/(4 \cdot n_{ARC})$, with λ_0 defining the free space wavelength [1]. The ARC is therefore usually designed to present the minimum at around 600 nm, where the flux of photons is a maximum in the solar spectrum. For the preparation of heterojunction solar cells in this thesis, TCO films are deposited on top of the amorphous or micro-crystalline emitter by DC magnetron sputtering from a ceramic target. In particular, an indium tin oxide (ITO) serves as TCO. The transparent conductive oxide films are fabricated using a single chamber DC magnetron sputtering setup (*VON ARDENNE*). The substrate temperature and the plasma power are adjusted to 200 °C and 200 W, respectively. The heater is attached in a defined distance of a few cms above the samples, which are placed into a horizontal rotating table. A second horizontal rotating table which gives various possibilities to operate the plasma, such as a special de-signed shadow mask, a fully open, or fully closed shield. The target and the samples are separated during heat-up time by the fully-closed metal shield. The sputtering time is set to 1100 s, so that a thickness of 80 nm had been obtained (corresponds to a reflection minimum at 600 nm). In Fig. A.1 the ab-sorption, reflection, and transmission spectra of ITO films used in this work are displayed. Figure A.2 illustrates the sample temperature inside the evac-uated chamber as a function of the heater temperature.

A.3. Metal contacts

Metal grids are used at the front side to collect the distributedly photogen-erated carriers. The trade-off between transparency and series resistance re-quires metalization technologies able to produce very narrow but thick and highly conductive metal lines with a low contact resistance to Si [1]. Therefore, the front side metal grid are fabricated using photolithography and *electron-gun* evaporation. Using a coarser metalization technique implies higher shad-ing and resistance losses, and restrict the efficiency enhancement that could

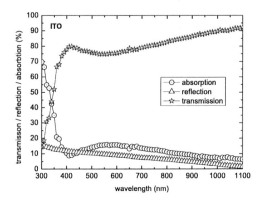

Figure A.1.: Absorption, reflection, and transmission spectra of ITO deposited with 200 W at 250 °C.

be achieved by internal cell design. Therefore, the laboratory cells used in this work are fabricated by means of photolithography and evaporation only, although these processes are not well suited to mass production.

To from the 4 μm metal fingers collecting the carriers from the TCO, a 6 μm photo-resist process is applied. The evaporated thickness is measured by a oscillating crystal. For the slightly n-type character ITO film, the combination of 30 nm Cr and 4 μm Ag combine low contact resistance and high bulk conductivity. The back-contact is made by full-side evaporation of Cr/Ag and Al for n-type and p-type substrates, respectively. The lift-off-method after the evaporation and a second photolithography process define the size of the laboratory type cells (1 by 1 cm or 2 by 2 cm) by ITO etching in diluted HF.

A.4. PECVD gases

A list of available gases at the PECVD setup used in this work in given in Table A.2. Those gaseous precursors are introduced to the PECVD chamber via mass-flow controllers (MFC), theirs maximal flow rate is denoted also in Table A.2.

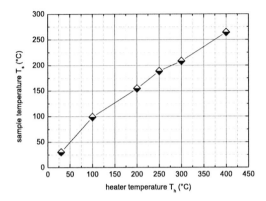

Figure A.2.: Sample temperature inside the evacuated chamber as a function of heater temperature.

Table A.2.: Available gases at the PECVD setup and corresponding installed mass flow controller (MFC) with minimum and maximum gas flow rate. It has to be noted that PH_3 is diluted 3 % in SiH_4, and TMB is diluted 2 % in H_2.

	PECVD chambers					
	n-type chamber		i-type chamber		p-type chamber	
channel	gas	MFC (sccm)	gas	MFC (sccm)	gas	MFC (sccm)
1	SiH_4	0 − 60	SiH_4	0 − 60	SiH_4	0 − 60
2	NH_3	0 − 146	H_2	0 − 200	TMB	0 − 10
3	H_2	0 − 200			CH_4	0 − 144
4	PH_3	0 − 12			SiH_4	0 − 30
5	N_2O	0 − 142			H_2	0 − 200
6	CH_4	0 − 144				
7	CO_2	0 − 145				

ABBREVIATIONS, ACRONYMS AND SYMBOLS

ΔE_c	conduction band discontinuity
ΔE_c	valence band discontinuity
Δn	injection level, excess carrier density
Δn_{av}	average excess carrier density
$\Delta n_{av}, \Delta p_{av}$	average excess carrier density
Φ_S	spectral photon flux density
Φ_a	absorbed spectral photon flux density
Φ_s	incident spectral photon flux density
Φ_{m1}, Φ_{m2}	work function
Ψ, Δ	related to the ratio of Fresnel reflection coefficients $\widetilde{R_p}$ and $\widetilde{R_s}$
α	absorption coefficient
χ	electron affinities
μc-Si:H	hydrogenated micro-crystalline silicon
μ_n, μ_p	electron and hole mobilities in c-Si
ν_{th}	thermal velocity of charge carriers
ψ	build-in voltage
σ_p, σ_n	capture cross sections
σ_{dark}	dark conductivity at room temperature
σ_{ph}	photoconductivity at room temperature
τ_{Auger}	Auger lifetime
τ_{SRH}	Shockley-Read-Hall lifetime
τ_{eff}	effective lifetime
τ_{bulk}	bulk recombination lifetime
τ_{p0}, τ_{n0}	electron and hole lifetimes related to the thermal velocity of charge carriers
$\tau_{radiative}$	radiative lifetime
$\tau_{surface}$	surface lifetime

$\widetilde{R_p}$ ($\widetilde{R_s}$)	Fresnel reflection coefficient for p- (s-) polarized light
μ-WPCD	microwave photoconductance decay
a_n, b_n	Fourier coefficients
a-Si	amorphous silicon
a-Si:H	hydrogenated amorphous silicon
a-SiC	amorphous silicon carbide
a-SiO	amorphous silicon oxide
a-SiO$_x$:H	hydrogenated amorphous silicon sub-oxides
ARC	anti reflection coating
B_2H_6	diborane
BSF	back surface field
c-Si	crystalline silicon
CO_2	carbon dioxide
D_n (D_p)	diffusion coefficient of electrons (holes)
D_{it}	interface state density
DRM	depletion region modulation
E_A	activation energy
E_c, E_v	conduction band and valence band energies
E_F	Fermi level
E_G	bandgap energy
E_t	thermal energy level of the defect
E_{04}	discrete bandgap
E_{Cody}	Cody bandgap
E_{opt}	optical bandgap
E_{ph}	photon energy
E_{Tauc}	Tauc bandgap
ECD	excess carrier density
EQE (IQE)	external (internal) quantum efficiency
H_2	hydrodgen
HF	hydrofluoric acid
HIT	heterojunction with intrinsic thin layer
I_{01}, I_{02}	current saturation
I_L	current generated by light
I_{sc}	short-circuit current

IPA	Isopropyl alcohol
ITO	Indium Tin oxide
J_0	saturation current density
J_{0BSF}	BSF saturation current density
J_{0E}	emitter saturation current density
J_{dark}	dark current density
J_{sc}	short circuit current density
L_n (L_p)	diffusion length of electrons (holes)
MCD	mean carrier density
MFC	mass-flow controller
n_0	electron concentration at thermal equilibrium
N_c, N_v	effective density of states at the conduction and valence band edges
n_i	intrinsic carrier concentration
N_T	density of recombination defects in the bulk
N_A	acceptor concentration
N_{dop}	semiconductor dopant density
N_D	donor concentration
n_{id1}, n_{id2}	diode ideality factor
N_{ts}	density of recombination defects at the surface
P	Polarizer azimuthal angle
PECVD	plasma enhanced chemical vapor deposition
PH_3	phosphine
q	electronic charge
QSSPC	quasi-steady-state photoconductance
R	reflectivity
R_s	series resistance
R_{sh} (R_p)	shunt (parallel) resistance
RCA	denotes a standard two-step cleaning process developed at Radio Corporation of America, USA
RT	room temperature
S	surface recombination velocity
S_{eff}	effective surface recombination velocity
S_{front} (S_{back})	surface recombination velocity at the front side (backside)

S_{n0} (S_{p0})	surface recombination velocity parameter for electrons (holes)
sccm	cubic centimeter per minute at STP
SE	spectroscopic ellipsometry
Si	silicon
SiH_4	silane
SiN_x	silicon nitride
SiO_2	silicon dioxide
SRH	Shockley-Read-Hall
SRV	surface recombination velocity
TCO	transparent conductive oxide
TLM	transfer length method
TMAH	Tetramethylammonium hydroxide
TMB	trimethylborane (C_3H_9B)
TPC	transient photoconductance
U_s	surface recombination rate
U_{eff}	effective recombination rate
U_{auger}	Auger recombination rate
U_{bulk}	bulk recombination rate
$U_{emitter}$	emitter recombination rate
$U_{radiative}$	radiative recombination rate
U_{SRH}	Shockley-Read-Hall recombination rate
$U_{surface}$	surface recombination rate
V_{oc}	open circuit voltage
VASE	variable angle spectral ellipsometer
VHF	very-high frequency
W	wafer thickness
W_{D1}, W_{D2}	comprehensiveness of the space charge region
W_{dpl}	width of space charge region

LIST OF FIGURES

LIST OF TABLES

BIBLIOGRAPHY

[1] A. LUQUE AND S. HEGEDUS, eds. Handbook of Photovoltaic Science and Engineering. Wiley, New York, USA (2003).

[2] C. HONSBERG AND S. BOWDEN. *Photovoltaics CDROM.* Tech. rep., University of Delaware (Solar Hydrogen IGERT) (2007).

[3] M. KAKU. Physics of the impossible. Doubleday Broadway Publishing Group, New York, USA (2008).

[4] S. W. GLUNZ. *High-Efficiency Crystalline Silicon Solar Cells.* Advances in OptoElectronics, vol. Article ID 97370, pp. 1–15 (2007).

[5] D. IDE, M. TAGUCHI, Y. YOSHIMINE, T. BABA, T. KINOSHITA, H. KANNO, H. SAKATA, E. MARUYAMA, AND M. TANAKA. *Excellent power-generating properties by using the HIT structure.* Proceedings of the 33rd IEEE Photovoltaic Specialists Conference (IEEE-PVSC-33), San Diego, USA (2008).

[6] M. TAGUCHI, K. KAWAMOTO, S. TSUGE, T. BABA, H. SAKATA, M. MORIZANE, K. UCHIHASHI, N. NAKAMURA, S. KIYAMA, AND O. OOTA. *HIT cells - high efficiency crystalline Si cells with novel structure.* Progress in Photovoltaics: Research and Applications, vol. 8, p. 503 (2000).

[7] T. SAWADA, N. TERADA, S. TSUGE, T. BABA, T. TAKAHAMA, K. WAKISAKA, S. TSUDA, AND S. NAKANO. *High-efficiency a-Si/c-Si heterojunction solar cell.* Proceedings of the 24th IEEE Photovoltaic Specialists Conference (IEEE-PVSC-24), Waikoloa, Hawaii, USA, p. 1219 (1994).

[8] M. TANAKA, M. TAGUCHI, T. MATSUYAMA, T. SAWADA, S. TSUDA, S. NAKANO, H. HANAFUSA, AND Y. KUWANO. *Development of new a-Si/c-Si heterojunction solar cells: ACJ-HIT (artificially constructed junction-heterojunction with intrinsic thin-layer).* Japanese Journal of Applied Physics (JJAP), vol. 31, p. 3518 (1992).

[9] M. TAGUCHI, A. TERAKAWA, E. MARUYAMA, AND M. TANAKA. *Obtaining a higher V_{oc} in HIT cells.* Progress in Photovoltaics: Research and Applications, vol. 13, p. 481 (2005).

[10] T. MUELLER, W. DUENGEN, Y. MA, R. JOB, M. SCHERFF, AND W. R. FAHRNER. *Investigation of the emitter band gap widening of heterojunction solar cells by use of hydrogenated amorphous carbon silicon alloys.* Journal of Applied Physics, vol. 102, p. 074505 (2007).

[11] T. MUELLER, S. SCHWERTHEIM, M. SCHERFF, AND W. R. FAHRNER. *High quality passivation for heterojunction solar cells by hydrogenated amorphous silicon suboxide films.* Applied Physics Letters, vol. 92, p. 033504 (2008).

[12] H. FUJIWARA, T. KANEKO, AND M. KONDO. *Application of hydrogenated amorphous silicon oxide layers to c-Si heterojunction solar cells.* Applied Physics Letters, vol. 91, p. 508 (2007).

[13] R. STANGL, A. FROITZHEIM, M. SCHMIDT, AND W. FUHS. *Design criteria for amorphous/crystalline silicon heterojunction solar cells - a simulation study.* Proceedings of the 3rd IEEE World Conference on Photovoltaic Energy Conversion (WCPEC-3), Osaka, Japan (2003).

[14] P. J. ROSTAN, U. RAU, AND J. H. WERNER. *TCO/(n-type)a-Si:H/(p-type)c-Si heterojunction solar cells with high open circuit voltage.* Proceedings of the 21th European Photovoltaic Solar Energy Conference and Exhibition (EU-PVSEC-21), Dresden, Germany (2006).

[15] S. OLIBET, E. VALLAT-SAUVAIN, AND C. BALLIF. *Model for a-Si:H/c-Si interface recombination based on the amphoteric nature of silicon dangling bonds.* Physical Review B, vol. 76, pp. 3261–32614 (2007).

[16] Q. WANG, M. R. PAGE, E. IWANCIZKO, Y. XU, L. ROYBAL, R. BAUER, D. LEVI, Y. YAN, T. WANG, AND H. M. BRANZ. *High open-circuit voltage in silicon heterojunction solar cells.* Materials Research Society Symposium A Proceedings, 2006 MRS Fall Meeting, Boston, USA (2007).

[17] M. TUCCI, L. SERENELLI, S. DE IULIIS, D. CAPUTO, A. NASCETTI, AND G. DE CESARE. *Amorphous/crystalline silicon heterostructure solar cells based on multi-crystalline silicon.* Proceedings of the 21st European Photovoltaic Solar Energy Conference and Exhibition (EU-PVSEC-21), Dresden, Germany (2008).

[18] H. FUJIWARA AND M. KONDO. *Effects of a-Si:H layer thickness on the performance of a-Si:H/c-Si heterojunction solar cells.* Journal of Applied Physics, vol. 101, p. 054516 (2007).

[19] Y. HAMAKAWA, K. FUJIMOTO, K. OKUDA, Y. KASHIMA, S. NONOMURA, AND H. OKAMOTO. *New types of high efficiency solar cells based on a-Si.* Applied Physics Letters, vol. 43, pp. 644–646 (1983).

[20] B. VON ROEDERN. *High-efficiency Si solar cell processing requirement: why 0.5 + 0.5 ≠ 1 (i.e., ≥3).* NREL 17th Workshop on Crystalline Silicon Solar Cells and Modules: Materials and Processes, Vail, Colorada USA, pp. 116–120 (2007).

[21] T. H. WANG, M. R. PAGE, E. IWANICZKO, Y. XU, Y. YAN, L. ROYBAL, D. LEVI, R. BAUER, H. M. BRANZ, AND Q. WANGA. *High-efficiency p-type silicon heterojunction solar cells.* Proceedings of the 21st European Photovoltaic Solar Energy Conference and Exhibition (EU-PVSEC-21), Dresden, Germany, pp. 781–783 (2006).

[22] H. KANNO, D. IDE, Y. TSUNOMURA, S. TAIRA, T. BABA, Y. YOSHIMINE, M. TAGUCHI, T. KINOSHITA, H. SAKATA, AND E. MARUYAMA. *Over 22 % Efficient HIT Solar Cell.* Proceedings of the 23rd European Photovoltaic Solar Energy Conference and Exhibition (EU-PVSEC-23), Valencia, Spain (2008).

[23] J. D. JOANNOPOULOS, G. LUCOVSKY, J. C. KNIGHTS, W. E. SPEAR, P. G. LECOMBER, M. J. THOMPSON, D. KAPLAN, D. E. CARLSON, AND A. MADAN. The Physics of Hydrogenated Amorphous Silicon I. Springer Verlag (1984).

[24] M. A. GREEN. High Efficiency Silicon Solar Cells. Trans Tech Publications (1987).

[25] W. P. MULLIGAN, D. H. ROSE, M. J. CUDZINOVIC, D. M. DeCEUSTER, K. R. McINTOSH, D. D. SMITH, AND R. M. SWANSON. *Manufacture of solar cells with 21% efficiency*. Proceedings of the 19th European Photovoltaic Solar Energy Conference and Exhibition (EU-PVSEC-19), Paris, France, pp. 387–390 (2004).

[26] M. J. KERR AND A. CUEVAS. *Very low bulk and surface recombination in oxidized silicon wafers*. Semiconductor Science and Technology, vol. 17, pp. 35–38 (2002).

[27] M. J. KERR AND A. CUEVAS. *Recombination at the interface between silicon and stoichiometric plasma silicon nitride*. Semiconductor Science and Technology, vol. 17, pp. 166–172 (2002).

[28] J. SCHMIDT. *Untersuchungen zur Ladungstraegerrekombination an den Oberflaechen und im Volumen von kristallinen Silicium-Solarzellen*. Ph.D. thesis, Universitaet Hannover (1998).

[29] I. MARTIN, M. VETTER, A. QORPELLA, J. PUIGDOLLERS, A. CUEVAS, AND R. ALCUBILLA. *Surface passivation of p-type crystalline Si by plasma enhanced chemical vapor deposited amorphous SiC$_x$:H films*. Applied Physics Letters, vol. 79, p. 2199 (2001).

[30] P. WUERFEL. Physik der Solarzellen. Spektrum Akademischer Verlag (2000).

[31] J. NELSON. The Physics of Solar Cells. Imperial College Press, UK (2003).

[32] M. A. GREEN. Operating Principles, Technology and System Applications. Centre for Photovoltaic Devices and Systems, University of South Wales, Sydney (1982).

[33] M. A. GREEN. Silicon Solar Cells - Advanced Principles and Practice. Centre for Photovoltaic Devices and Systems, University of South Wales, Sydney (1995).

[34] A. GOETZBERGER, J. KNOBLOCH, AND B. VOSS. Crystalline Silicon Solar Cells. Wiley, New York, USA (1998).

[35] W. VAN ROOSBROECK. *Theory of the flow of electrons and holes.* Bell System Technical Journal, vol. 29, pp. 560–607 (1950).

[36] S. M. SZE. Physics of Semiconductor Devices. John Wiley and Sons, N.Y. (1981).

[37] R. HAECKER AND A. HANGLEITER. *Intrinsic upper limits of the carrier lifetime in silicon.* Journal of Applied Physics, vol. 75, p. 7570 (1994).

[38] M. J. KERR. *Surface, Emitter and Bulk Recombination in Silicon and Development of Silicon Nitride Passivated Solar Cells.* Ph.D. thesis, Australien National University (2002).

[39] P. T. LANDSBERG. Recombination in Semiconductors. Cambridge University Press, New York (1991).

[40] M. SHUR. Physics of Semiconductor Devices. Englewood Cliffs, Prentice Hall (1990).

[41] M. S. TYAGI. Introduction to Semiconductor Materials and Devices. Chichester, Wiley, New York, USA (1991).

[42] A. HANGLEITER AND R. HACKER. *Enhancement of band-to-band Auger recombination by electron-hole correlations.* Physical Review Letters, vol. 65, pp. 215–218 (1990).

[43] J. DZIEWIOR AND W. SCHMID. *Auger coefficients for highly doped and highly excited silcion.* Applied Physics Letters, vol. 31, pp. 346–348 (1977).

[44] J. D. BECK AND R. CONRADT. *Auger-recombination in Si.* Solid State Communications, vol. 13, p. 93 (1973).

[45] A. G. ABERLE. Crystalline Silicon Solar Cells - Advanced Surface Passivation and Analysis. Univeristy of New South Wales, Sydney (1999).

[46] P. P. ALTERMATT, J. SCHMIDT, G. HEISER, AND A. G. ABERLE. *Assessment and parameterisation of Coloumb-enhanced Auger recombination coefficients in lowly injected crystalline silicon.* Journal of Applied Physics, vol. 82, p. 4938 (1997).

[47] W. SHOCKLEY AND W. T. READ-JR. *Statistics of the recombination of holes and electrons.* Physical Review, vol. 87, pp. 835–842 (1952).

[48] R. N. HALL. *Electron-hole recombination in germanium.* Physical Review, vol. 87, p. 387 (1952).

[49] K. L. LUKE AND L. CHENG. *Analysis of the interaction of a laser pulse with a silicon wafer: Determination of bulk lifetime and surface recombination velocity.* Journal of Applied Physics, vol. 61, pp. 2282–2293 (1987).

[50] W. D. EADES. Electrical Engineering, chap. Characterization of Silicon-Silicon Dioxide Interface Traps using Deep Level Transient Spectroscopy, p. 137. Stanford (1985).

[51] J. MANDELKORN AND J. H. LAMNECK. *Simplified fabrication of back surface electric field silicon cells and novel characteristics of such cells.* Proceedings of the 9th IEEE Photovoltaic Specialists Conference (IEEE-PVSC-9), Silver Springs, USA (1972).

[52] D. E. KANE AND R. M. SWANSON. *Measurement of the emitter saturation current by a contactless photoconductivity decay method.* Proceedings of the 18th IEEE Photovoltaic Specialists Conference (IEEE-PVSC-18), Las Vegas, USA, pp. 578–583 (1985).

[53] F. A. LINDHOLM AND C. T. SAH. *Normal modes of semiconductor p-n junction devices for material-parameter determination.* Journal of Applied Physics, vol. 47, p. 4203 (1976).

[54] A. NEUGROSCHEL, F. A. LINDHOLM, AND C. T. SAH. *A method for determining the emitter and base lifetimes in p-n junction diodes.* IEEE Transactions on Electron Devices, vol. 24, p. 662 (1977).

[55] S. C. JAIN AND R. MURALIDHAN. *Effect of emitter recombination on the open circuit voltage decay of a junction diode.* Solid State Electronics, vol. 24, p. 1147 (1981).

[56] F. A. LINDHOLM, A. NEUGROSCHEL, C. T. SAH, M. D. GODLEWSKI, AND H. W. BRANDHORST JR. *A methodology for experimentally based determination of gap shrinkage and effective lifetimes in the emitter and base of p-n junction solar cells and other p-n junction devices.* IEEE Transactions on Electron Devices, vol. 24, p. 402 (1977).

[57] A. NEUGROSCHEL, P. J. CHEN, S. C. PAO, AND F. A. LINDHOLM. *Diffusion length and lifetime determination in p-n junction solar cells and diodes by forward-biased capacitance measurements.* IEEE Transactions on Electron Devices, vol. 25, p. 485 (1978).

[58] D. KRAY. *Hocheffiziente Solarzellenstrukturen fuer kristallines Silicium-Material industrieller Qualitaet.* Ph.D. thesis, Universitaet Konstanz, ISE Freiburg (2004).

[59] S. W. GLUNZ. *Ladungstraegerrekombination in Silizium und Siliziumsolarzellen.* Ph.D. thesis, Universitaet Freiburg (1995).

[60] R. A. STREET. Hydrogenated amorphous silicon. Cambridge University Press (1991).

[61] R. JANSSEN, A. JANOTTA, D. DIMOVA-MALINOVSKA, AND M. STUTZMANN. *Optical and electrical properties of doped amorphous silicon suboxides.* Physical Review B, vol. 60, p. 13561 (1999).

[62] J. TAUC. Amorphous and liquid semiconductors. Plenum Press (1974).

[63] E. A. DAVIS AND N. F. MOTT. *Conduction in non-crystalline systems V: Conductivity, optical absorption and photoconductivity in amorphous semiconductors.* Philosophical Magazine, vol. 22:179, pp. 903–922 (1970).

[64] J. KANICKI, ed. Amorphous & Microcrystalline Semiconductor Devices II - Materials and Device Physics. Artech House, Boston (1992).

[65] R. ANDERSON. *Experiments on Ge-GaAs heterojunctions.* Solid State Elec-
      tronics, vol. 5, p. 341 (1962).

[66] S. M. SZE AND K. K. NG. Physics of Semiconductor Devices. Wiley-
      Interscience (2007).

[67] A. G. MILNES AND D. L. FEUCHT. Heterojunctions and Metal Semiconduc-
      tor Junctions. Academic Press, New York (1972).

[68] B. L. SHARMA AND P. K. PUROHIT. Semiconductor Heterojunctions. Perg-
      amon, Oxford (1974).

[69] W. R. FRENSLEY AND H. KROEMER. *Theory of the energy-band lineup at
      an abrupt semiconductor heterojunction.* Physical Review B, vol. 16, p.
      2642 (1977).

[70] M. J. ADAM AND A. NUSSBAUM. *A proposal for a new approach to hetero-
      junction theory.* Solid State Electronics, vol. 22, p. 783 (1979).

[71] O. VON ROSS. *Theory of extrinsic and intrinsic heterojunctions in thermal
      equilibrium.* Solid State Electronics, vol. 23, p. 1069 (1980).

[72] W. A. HARRISON. *Elementary theory of heterojunctions.* Journal of Vac-
      uum Science and Technology, vol. 14, p. 1016 (1977).

[73] J. TERSOFF. *Theory of semiconductor heterojunctions: The role of quan-
      tum dipoles.* Physical Review B, vol. 30, p. 4874 (1984).

[74] R. PREU. *Innovative Produktionstechnologien fuer kristalline Silicium-
      Solarzellen.* Ph.D. thesis, University of Hagen, LGBE (2000).

[75] H. MELCHIOR. Laser Handbook, Volume 1, chap. Demodulation and
      Photodetection Techniques, pp. 725–835. F. T. Arecci and E. O. Schulz-
      Dubois, Amsterdam (1972).

[76] G. D. CODY, B. ABELES, C. WRONSKI, C. R. STEPHENS, AND B. BROOKS.
      Optical characterization of amorphous silicon-hybride films. Solar Cells,
      vol. 2, p. 227 (1980).

[77] O. SCHULTZ. *High-Efficiency Multicrystalline Silicon Solar Cells.* Ph.D. thesis, Universitaet Konstanz (2005).

[78] R. M. SWANSON. *Approaching the 29 % limit efficiency of silicon solar cells.* Proceedings of the 31st IEEE Photovoltaic Specialists Conference (IEEE-PVSC-31), Lake Buena Vista, USA (2005).

[79] M. TAGUCHI. *Improvment of the conversion efficiency of polycrystalline silicon thin film solar cells.* Proceedings of the 5th International Photovoltaic Science and Engineering Conference (Asia-PVSEC-90), Kyoto, Japan, pp. 689–692 (1990).

[80] K. WAKISAKA, M. TAGUCHI, T. SAWADA, M. TANAKA, T. MATSUYAMA, T. MATSUOKA, S. TSUDA, S. NAKANO, Y. KISHI, AND Y. KUWANO. *More than 16 % solar cells with a new HIT (doped a-Si / non-doped a-Si / crystalline Si) structure.* Proceedings of the 22nd IEEE Photovoltaic Specialists Conference (IEEE-PVSC-22), Las Vegas, USA (1991).

[81] M. PAGE, E. IWANICZKO, Y. XU, Q. WANG, Y. YAN, L. ROYBAL, H. M. BRANZ, AND T. WANG. *Well passivated a-Si:H Back Contacts for Double-HeterojunctionSilicion Solar Cells.* Proceedings of the 4th IEEE World Conference on Photovoltaic Energy Conversion (WCPEC-4), Waikoloa, Hawaii, USA, pp. 1485–1488 (2006).

[82] E. MARUYAMA, A. TERAKAWA, M. TAGUCHI, Y. YOSHIMINE, D. IDE, T. BABA, M. SHIMA, H. SAKATA, AND M. TANAKA. *Sanyo's challenges to the development of high-efficiency HIT solar cells and the expansion of HIT business.* Proceedings of the 4th IEEE World Conference on Photovoltaic Energy Conversion (WCPEC-4), Waikoloa, Hawaii, USA, p. 1455 (2006).

[83] S. TSUDA, T. TAKAHAMA, Y. HISHIKAWA, H. TARUI, K. NISHIWAKI, K. WAKISAKA, AND S. NAKANO. *a-Si technologies for high efficiency solar cells.* Journal of Non-Crystalline Solids, vol. 164-166, p. 679 (1993).

[84] K. KURITA AND T. SHINGYOUJI. *Low surface recombination velocity on*

silicon wafer surface due to Iodine-Ethanol treatment. Japanese Journal of Applied Physics (JJAP), vol. 38, pp. 5710–5714 (1999).

[85] M. WOLF. *High efficiency silicon solar cells.* Proceedings of the 14th IEEE Photovoltaic Specialists Conference (IEEE-PVSC-14), San Diego, USA, pp. 674–679 (1980).

[86] S. OLIBET, E. VALLAT-SAUVAIN, AND C. BALLIF. *Effect of light induced degradation on passivating properties of a-Si:H layers deposited crystalline Si.* Proceedings of the 21st European Photovoltaic Solar Energy Conference and Exhibition (EU-PVSEC-21), Dresden, Germany (2006).

[87] A. M. FROITZHEIM. *Hetero-Solarzellen aus amorphem und kritallinem Silizium.* Ph.D. thesis, Philipps-Universitaet Marburg (2003).

[88] G. GRABOSCH. *Herstellung und Charakterisierung von PECVD abgeschiedenen mikrokristallinem Silizium.* Ph.D. thesis, University of Hagen, LGBE (2000).

[89] M. A. LIEBERMAN AND A. J. LICHTENBERG. Principles of plasma discharges and materials processing. John Wiley & Sons, New York (1994).

[90] H. FUJIWARA AND M. KONDO. *Interface structure in a-Si:H/c-Si heterojunction solar cells characterized by optical diagnoses technique.* Proceedings of the 4th IEEE World Conference on Photovoltaic Energy Conversion (WCPEC-4), Waikoloa, Hawaii, USA, pp. 1443–1448 (2006).

[91] J. L. VOSSEN AND J. J. CUOMO. Thin Film Processes. New York: Academic (1978).

[92] P. R. I CABARROCAS. Properties of amorphous silicon and its alloys, chap. 1.1, p. 3. emis Datareviews series no. 19, INSPEC (1989).

[93] H. G. TOMPKINS AND W. A. McGAHAN. Spectroscopic Ellipsometry and Reflectometry: A User's Guide. John Wiley and Sons, Inc. (1999).

[94] WOOLAM. Guiding to Using WVASE32, Software for VASE Ellipsometers. J. A. Woolam Co., Inc. (1997).

[95] G. E. JELLISON-JR. AND F. A. MODINE. *Parameterization of the optical functions of amosphous materials in the interband region.* Applied Physics Letters, vol. 69, p. 371 (1996).

[96] Y. HAMAKAWA, Y. TAWADA, K. NISHIMURA, K. TSUGE, M. KONDO, K. FUJIMOTO, S. NONOMURA, AND H. OKAMOTO. *Design parameters of high efficiency a-SiC:H/a-Si:H heterojunction solar cells.* Proceedings of the 16th IEEE Photovoltaic Specialists Conference (IEEE-PVSC-16), San Diego, USA (1982).

[97] G. D. CODY, B. G. BROOKS, AND B. ABELES. *Optical absorption above the optical gap of amorphous silicon hybride.* Solar Energy Materials, vol. 8, pp. 231–240 (1982).

[98] G. D. CODY. Semiconductors and Semimetals, vol. 21B, p. 11. Academic, New York (1984).

[99] J. R. FERRARO AND K. NAKAMOTO. Introductory Raman Spectroscopy. Academic Press (1994).

[100] W. SUETAKA. Surface Infrared and Raman Spectroscopy. Plenum Press, New York (1995).

[101] G. TURRELL AND J. CORSET, eds. Raman Microscopy - Developments and Applications. Academic Press (1996).

[102] B. SCHRADER, ed. Infrared and Raman Spectroscopy. VCH (1995).

[103] J. J. LASERNA, ed. Modern Techniques in Raman Spectroscopy. John Wiley & Sons (1996).

[104] D. K. SCHRODER. Semiconductor Material and Device Characterization. John Wiley & Sons, New York, USA (1998).

[105] H. OVERHOF AND P. THOMAS. Electronic Transports in Hydrogenated Amorphous Semiconductors. Springer, Berlin (1989).

[106] C. VOICU. *Entwicklung und Aufbau einer Steuerung für einen Solarsim-
ulator*. Master's thesis, Lehrstuhl für Bauelemente der Elektrotechnik,
FernUniversitaet Hagen (2003).

[107] R. A. SINTON. WCT-120 Photoconductance Lifetime Tester and optional
Suns-Voc Stage. Sinton Consulting, Inc, 1132 Green Circle Boulder, CO
80305 USA (2006). Revised by Daniel MacDonald.

[108] P. A. BASORE AND D. A. CLUGSTON. *PC1D version 4 for windows: from
analysis to design*. Proceedings of the 23rd IEEE Photovoltaic Specialists
Conference (IEEE-PVSC-33), Louisville, USA, p. 377 (1993).

[109] M. KUNST AND G. BECK. *The study of charge carrier kinetics in semiconductors by microwave conductivity measurements*. Journal of Applied
Physics, vol. 60, p. 3558 (1986).

[110] M. KUNST AND G. BECK. *The study of charge carrier kinetics in semiconductors by microwave conductivity measurements. II.* Journal of Applied
Physics, vol. 63, p. 1093 (1987).

[111] P. A. BASORE AND B. R. HANSEN. *Microwave-detected photoconductance
decay*. Proceedings of the 21st IEEE Photovoltaic Specialists Conference
(IEEE-PVSC-21), Orlando, USA, p. 374 (1990).

[112] R. A. SINTON, A. CUEVAS, AND M. STUCKINGS. *Quasi-steady-state photoconductance, a new method for solar cell material and device characterization*. Proceedings of the 25th IEEE Photovoltaic Specialists Conference
(IEEE-PVSC-25), Washington, USA, pp. 457–460 (1996).

[113] R. A. SINTON AND A. CUEVAS. *Contactless determination of current—
voltage characteristics and minority carrier lifetimes in semiconductors
from quasi-steady-state photoconductance data*. Applied Physics Letters,
vol. 69, pp. 2510–2515 (1996).

[114] H. NAGEL, C. BERGE, AND A. G. ABERLE. *Generalized analysis of quasi-
steady-state and quasi-transient measurements of carrier lifetimes in
semiconductors*. Journal of Applied Physics, vol. 86, pp. 6218–6221
(1999).

[115] F. DANNHAEUSER. *Die Abhaengigkeit der Traegerbeweglichkeit in Silizium von der Konzentration der freien Ladungstraeger-I.* Solid State Electronics, vol. 15, pp. 1371–1375 (1972).

[116] J. KRAUSSE. *Die Abhaengigkeit der Traegerbeweglichkeit in Silizium von der Konzentration der freien Ladungstraeger-II.* Solid State Electronics, vol. 15, pp. 1377–1381 (1972).

[117] P. A. BASORE AND D. A. CLUGSTON. *PC1D version 5: 32-bit solar cell modeling on personal computers.* Proceedings of the 26th IEEE Photovoltaic Specialists Conference (IEEE-PVSC-26), Anaheim, USA, pp. 207–210 (1997).

[118] S. M. SZE. Semiconductor Devices: Physics and Technology. Cambridge University Press, Cambridge, 2nd edn. (2001).

[119] G. S. KOUSIK, Z. G. LING, AND P. K. AJMERA. *Nondestructive technique to measure bulk lifetime and surface recombination velocities at the two surfaces by infrared absorption due to pulsed optical excitation.* Journal of Applied Physics, vol. 72, p. 141 (1992).

[120] K. L. LUKE AND L. J. CHENG. *A chemical/microwave technique for the measurement of bulk minority carrier lifetime in silicon wafers.* Journal of The Electrochemical Society (JES), vol. 135, p. 957 (1987).

[121] J. SCHMIDT AND A. G. ABERLE. *Carrier recombination at silicon-silicon nitride interfaces fabricated by plasma-enhanced chemical vapor deposition.* Journal of Applied Physics, vol. 85, p. 3626 (1999).

[122] A. CUEVAS AND R. A. SINTON. *Prediction of the open-circuit voltage of solar cells from steady-state photoconductance.* Progress in Photovoltaics: Research and Applications, vol. 5, pp. 79–90 (1997).

[123] R. A. STREET. Technology and Applications of Amorphous Silicon. Springer (2000).

[124] Y. HAMAKAWA. *Present status of solar photovoltaic R&D projects in Japan.* Surface Science, vol. 86, p. 444 (1979).

[125] H. OKAMOTO, T. YAMAGUCHI, AND Y. HAMAKAWA. *Effect of DC electric field on the basic properties of RF plasma deposited a-Si.* Journal of Non-Crystalline Solids, vol. 35-36, pp. 201–206 (1980).

[126] H. OKAMOTO, Y. NITTA, T. YAMAGUCHI, AND Y. HAMAKAWA. *Device physics and design of a-Si ITO/p-i-n heteroface solar cells.* Solar Energy Materials, vol. 2, p. 313 (1980).

[127] Y. NITTA, H. OKAMOTO, AND Y. HAMAKAWA. *Amorphous Si: heteroface photovoltaic cells based upon p-i-n junction structure.* Japanese Journal of Applied Physics - 1st Photovoltaic Science and Engineering Conf. Japan, Tokyo, 1979, vol. 19, pp. 143–148 (1980).

[128] M. CARDONA. *Vibrational spectra of hydrogen in silicon and germanium.* Physica Status Solidi (B) - Basic Research, vol. 118, p. 463 (1983).

[129] B. RACINE, A. C. FERRARI, N. A. MORRISON, I. HUTCHINGS, W. I. MILNE, AND J. ROBERTSON. *Properties of amorphous carbon–silicon alloys deposited by a high plasma density source.* Journal of Applied Physics, vol. 90, p. 5002 (2001).

[130] I. SOLOMON, M. P. SCHMIDT, AND H. TRAN-QUOC. *Selective low-power plasma decomposition of silane-methane mixtures for the preparation of methylated amorphous silicon.* Physical Review B, vol. 38, p. 9895 (1988).

[131] E. GOGOLIDES, D. MARY, A. RHALLABI, AND G. TURBAN. *RF plasmas in methane: prediction of plasma properties and neutral radical densities with combined gas-phase physics and chemistry model.* Japanese Journal of Applied Physics (JJAP), vol. 34, p. 261 (1995).

[132] W. BEYER. Hydrogen in Semiconductors II, vol. 61, chap. 2. Academic Press (San Diego) (1999).

[133] J. BULLOT AND M. P. SCHMIDT. *Physics of amorphous silicon-carbon alloys.* Physica Status Solidi (B) - Basic Research, vol. 143, p. 345 (1987).

[134] B. A. KOREVAAR, J. FRONHEISER, X. ZHANG, L. M. FEDOR, AND T. R. TOLLIVER. *Influence of annealing on performance for heterojunction a-Si/c-Si*

devices. Proceedings of the 33rd IEEE Photovoltaic Specialists Conference (IEEE-PVSC-33), San Diego, USA (2008).

[135] M. NISHIDA, T. SHINDO, Y. KOMATSU, S. OKAMOTO, M. KANEIWA, AND T. NANMORI. *Single-crystalline silicon solar cell with pp^+ heterojunction of c-Si substrate and μc-Si : H film.* Solar Energy Materials and Solar Cell, vol. 48, p. 131 (1997).

[136] H. D. GOLDBACH, A. BINK, AND R. E. I. SCHROPP. *Thin p^{++} μc-Si layers for use as back surface field in p-type silicon heterojunction solar cells.* Journal of Non-Crystalline Solids, vol. 352, p. 1872 (2006).

[137] R. KLIMKEIT. *Herstellung von hochdotierten mikrokristallinen hydrogenisierten Siliziumschichten (n^+/p^+ μc-Si:H) mittels PECVD-Verfahren).* Master's thesis, University of Hagen, LGBE (2008).

[138] J. K. RATH, C. H. M. VAN DER WERF, F. A. RUBINELLI, AND R. E. I. SCHROPP. *Development of amorphous silicon based p-i-n solar cell in asuperstrate structure with p-microcrystalline silicon as window layer.* Proceedings of the 25th IEEE Photovoltaic Specialists Conference (IEEE-PVSC-25), Washington, DC, USA (1996).

[139] B. RECH. *Solarzellen aus amorphem Silizium mit hohem stabilem Wirkungsgrad.* Ph.D. thesis, RWTH Aachen, Germany (1997).

[140] M. ROESCH. *Experimente und numerische Simulation zum Ladungstraegertransport in a-Si:H/c-Si Heterodioden.* Ph.D. thesis, Universitaet Oldenburg (2003).

[141] A. LAADES, K. KLIEFOTH, L. KORTE, K. BRENDEL, R. STANGL, M. SCHMIDT, AND W. FUHS. *Surface passivation of crystalline silicon wafers by hydrogenated amorphous silicon probed by time resolved surface photovoltage and photoluminescence spectroscopy.* Proceedings of the 19th European Photovoltaic Solar Energy Conference and Exhibition (EU-PVSEC-19), Paris, France, pp. 1170–1173 (2004).

[142] S. D. WOLF AND G. BEAUCARNE. *Surface passivation properties of boron-doped plasma-enhanced chemical vapor deposited hydrogenated amorphous silicon films on p-type crystalline Si substrates.* Journal of Applied Physics, vol. 88, p. 022104 (2006).

[143] S. DAUWE, J. SCHMIDT, AND R. HEZEL. *Very low surface recombination velocities on p- and n-type silicon wafers passivated with hydrogenated amorphous silicon films.* Proceedings of the 29th IEEE Photovoltaic Specialists Conference (IEEE-PVSC-29), New Orleans, USA, pp. 1246–1249 (2002).

[144] T. WANG, E. IWANICZKO, M. PAGE, D. LEVI, Y. YAN, V. YELUNDUR, H. BRANZ, A. ROHATGI, AND Q. WANG. *Effective Interfaces in Silicon Heterojunction Solar Cells.* Proceedings of the 31st IEEE Photovoltaic Specialists Conference (IEEE-PVSC-31), Orlando, USA (2005).

[145] M. VETTER, I. MARTIN, R. FERRE, M. GARIN, AND R. ALCUBILLA. *Crystalline silicon surface passivation by amorphous silicon carbide films.* Solar Energy Materials and Solar Cells, vol. 91, pp. 174–179 (2006).

[146] M. GARIN, U. RAU, W. BRENDLE, I. MARTIN, AND R. ALCUBILLA. *Characterization of a-Si:H/c-Si interfaces by effective-lifetime measurements.* Journal of Applied Physics, vol. 98, p. 093711 (2005).

[147] M. VETTER, Y. TOUATI, I. MARTIN, R. FERRE, R. ALCUBILLA, I. TORRES, J. ALONSO, AND M. VAZQUEZ. *Characterization of industrial p-type CZ silicon wafers passivated with a-SiC$_x$:H films.* Proceedings of the 2005 IEEE Spanish Conference on Electronic Devices (CDE05), Tarragona, Spain, pp. 247–250 (2005).

[148] T. MUELLER, W. DUENGEN, R. JOB, M. SCHERFF, AND W. R. FAHRNER. *Crystalline silicon surface passivation by PECV deposited hydrogenated amorphous silicon oxide Films (a-SiO$_x$:H).* Materials Research Society Symposium A Proceedings, 2006 MRS Fall Meeting, Boston, USA, pp. 0989–A05–02 (2007).

[149] D. DAS, S. IFTIQUAR, AND A. BARUA. *Wide optical-gap a-SiO:H films prepared by rf glow discharge.* Journal of Non-Crystalline Solids, vol. 210, pp. 148–154 (1997).

[150] A. JANOTTA, R. JANSSEN, M. SCHMIDT, T. GRAF, M. STUTZMANN, L. GÖRGENS, A. BERGMAIER, G. DOLLINGER, C. HAMMERL, S. SCHREIBER, AND B. STRITZKER. *Doping and its efficiency in a-SiO_x:H.* Physical Review B, vol. 69, p. 115206 (2004).

[151] H. YAMAMOTO, M. OHTSUBO, T. SUGIURA, A. LIMMANEE, T. SATO, S. MIYAJIMA, A. YAMADA, AND M. KONAGAI. *Low temperature deposition of a-SiO:H films for high quality rear surface passivation.* Proceedings of the 22nd European Photovoltaic Solar Energy Conference and Exhibition (EU-PVSEC-22), Milan, Germany (2007).

[152] J. SRITHARATHIKHUN, C. BANERJEE, M. OTSUBO, T. SUGIURA, H. YAMAMOTO, T. SATO, A. LIMMANEE, A. YAMADA, AND M. KONAGAI. *Surface passivation of crystalline and polycrystalline silicon using hydrogenated amorphous silicon oxide film.* Japanese Journal of Applied Physics (JJAP), vol. 46, pp. 3296–3300 (2007).

[153] D. K. BIEGELSEN, R. A. STREET, C. C. TSAI, AND J. C. KNIGHTS. *Hydrogen evolution and defect creation in amorphous Si: H alloys.* Physical Review B, vol. 20, p. 4839 (1979).

[154] B. HOEX, F. J. J. PEETERS, M. CREATORE, M. A. BLAUW, W. M. M. KESSELS, AND M. C. M. VAN DE SANDEN. *High-rate plasma-deposited SiO_2 films for surface passivation of crystalline silicon.* Journal of Vacuum Science & Technology A, vol. 24 (5), p. 1823 (2006).

[155] J. D. KUBICKI AND D. SYKES. *Molecular orbital calculations on $H_6Si_2O_7$ with ; variable Si-O-Si angle: implications for the high-pressure vibrational spectra of silicate glasses.* American Mineralogist, vol. 78, pp. 253–259 (1993).

[156] M. PARK, C. W. TENG, V. SAKHRANI, AND M. B. MCLAURIN. *Optical characterization of wide band gap amorphous semiconductors (a-SiC:H): Effect*

of hydrogen dilution. Journal of Applied Physics, vol. 89, pp. 1130–1137 (2001).

[157] E. YABLONOVITCH AND T. GMITTER. *Auger recombination in silicon at low carrier densities.* Applied Physics Letters, vol. 49, p. 587 (1986).

[158] S. OLIBET, E. VALLAT-SAUVAIN, C. BALLIF, L. KORTE, AND L. FRESQUET. *Silicon solar cell passivation using heterostructures.* NREL 17th Workshop on Crystalline Silicon Solar Cells and Modules: Materials and Processes, Vail, Colorada USA (2007).

[159] S. OLIBET, C. MONACHON, A. HESSLER-WYSER, E. VALLAT-SAUVAIN, S. D. WOLF, L. FESQUET, J. DAMON-LACOSTE, AND C. BALLIF. *Textured silicon heterojunction solar cells with over 700 mV open-circuit voltage studied by transmission electron microscopy.* Proceedings of the 23nd European Photovoltaic Solar Energy Conference and Exhibition (EU-PVSEC-23), Valencia, Spain (2008).

[160] A. MATSUDA AND K. TANAKA. *Guiding principle for preparing highly photosensitve Si-based amorphous alloys.* Journal of Non-Crystalline Solids, vol. 97, p. 1367 (1987).

[161] S. NONOMURA, N. YOSHIDA, AND T. ITOH. *The formation of heterojunction using carbon alloys by hot-wire CVD method.* Thin Solid Films, vol. 501, pp. 164–168 (2006).

[162] T. HATTORI. Ultraclean Surface Processing of Silicon Wafers. Springer, New York (1998).

[163] N. JENSEN. *Heterostruktursolarzellen aus amorphem und kristallinem Silicium.* Ph.D. thesis, Universitaet Stuttgart (2002).

[164] H. ANGERMANN, W. HENRION, M. REBIEN, AND A. ROESELER. *Wet-chemical preparation and spectroscopic characterization of Si interfaces.* Applied Surface Science, vol. 235, p. 322 (2004).

LIST OF PUBLICATIONS

This thesis is a monograph, which contains some unpublished material, but is mainly based on the following publications including refereed journal paper and refereed paper presented at international conferences:

[1] T. MUELLER, S. SCHWERTHEIM, K. MEUSINGER, B. WDOWIAK, R. KLIMKEIT, AND W. R. FAHRNER. *Application of Plasma Deposited Nanocomposite Silicon Suboxides and Microcrystalline Silicon Alloys to Heterojunction Solar Cells.* Proceedings of the 34th IEEE Photovoltaic Specialists Conference (IEEE-PVSC-34), Philadelphia, USA (2009).

[2] T. MUELLER, S. SCHWERTHEIM, AND W. R. FAHRNER. *Application of wide-bandgap hydrogenated amorphous silicon oxide layers to heterojunction solar cells for high quality passivation.* Proceedings of the 33rd IEEE Photovoltaic Specialists Conference (IEEE-PVSC-33), San Diego, USA (2008).

[3] T. MUELLER, S. SCHWERTHEIM, M. SCHERFF, AND W. R. FAHRNER. *High quality passivation for heterojunction solar cells by hydrogenated amorphous silicon suboxide films.* Applied Physics Letters, vol. 92, p. 033504 (2008).

[4] T. MUELLER, W. DUENGEN, R. JOB, M. SCHERFF, AND W. R. FAHRNER. *Crystalline silicon surface passivation by PECV deposited hydrogenated amorphous silicon oxide films (a-SiO$_x$:H).* Materials Research Society Symposium A Proceedings, 2006 MRS Fall Meeting, Boston, USA, vol. 0989, pp. A05-02 (2007).

[5] T. MUELLER, W. DUENGEN, Y. MA, R. JOB, M. SCHERFF, AND W. R. FAHRNER. *Investigation of the emitter band gap widening of heterojunction solar cells by use of hydrogenated amorphous carbon silicon alloys.* Journal of Applied Physics, vol. 102, p. 074505 (2007).

[6] T. MUELLER, M. SCHERFF, S. SCHWERTHEIM, AND W. R. FARHNER. *Hydrogenated amorphous silicon sub-oxide films (a-SiO$_x$:H) for high quality crystalline silicon surface passivation.* Technical Digest of the 17th International Photovoltaic Science and Engineering Conference (Asia-PVSEC-17), Fukuoka, Japan, pp. 4P-P2-22, 717–718 (2007).

[7] T. MUELLER, S. SCHWERTHEIM, Y. HUANG, M. SCHERFF, AND W. R. FAHRNER. *Surface passivation of heterojunction solar cells using PECV deposited hydrogenated amorphous silicon oxide layers.* NREL 17th Workshop on Crystalline Silicon Solar Cells and Modules: Materials and Processes, Vail, Colorado USA. NREL/BK-520-41973 (2007).

[8] T. MUELLER, S. SCHWERTHEIM, M. SCHERFF, U. ZASTROW, Y. HUANG, AND W. R. FAHRNER. *Crystalline silicon surface passivation by PECV deposited hydrogenated amorphous silicon oxide layers.* Proceedings of the 22nd European Photovoltaic Solar Energy Conference and Exhibition (EU-PVSEC-22), Milan, Italy (2007).

[9] T. MUELLER, W. DUENGEN, Y. MA, M. SCHERFF, AND W. R. FAHRNER. *μ-Raman spectra analysis of hydrogenated amorphous carbon-silicon alloys (a-SiC(n):H) used as emitter window layers in hetero-junction solar cell structures.* Proceedings of the 21st European Photovoltaic Solar Energy Conference and Exhibition (EU-PVSEC-21), Dresden, Germany, pp. 1452–1455 (2006).

[10] T. MUELLER, M. SCHERFF, W. DUENGEN, Y. MA, AND W. R. FAHRNER. *Hydrogenated amorphous carbon-silicon alloys (a-SiC(n):H) used as emitters of heterojunction solar cells.* Proceedings of the 4th IEEE World Conference on Photovoltaic Energy Conversion (WCPEC-4), Waikoloa, Hawaii, USA, pp. 1407–1410 (2006).

[11] T. MUELLER, Y. HUANG, M. SCHERFF, AND W. R. FAHRNER. *Efficiency improvement of HIT solar cells by adding carbon in the emitter.* Proceedings of the 15th International Photovoltaic Science and Engineering Conference (Asia-PVSEC-15), Shanghai, China (2005).

OTHER RELATED PUBLICATIONS

[1] W. R. FAHRNER, M. SCHERFF, T. MUELLER, AND S. SCHWERTHEIM. Micromaterials and Nanomaterials, vol. 9, chap. Determination of Interface Defect Properties in a-Si:H/mc-Si Heterojunction Solar Cells, pp. 101–105. ISSN 1619-2486, Goldbogen Verlag, Dresden, Germany (2009).

[2] R. AUER, N. GAWEHNS, T. KUNZ, T. MUELLER, H. WINKELMANN, AND W. R. FAHRNER. *Crystalline Silicon Thin-Film Cells With Heterojunction Emitter.* Proceedings of the 23rd European Photovoltaic Solar Energy Conference and Exhibition (EU-PVSEC-23), Valencia, Spain, 3AV.1.25 (2008).

[3] R. AUER, N. GAWEHNS, T. KUNZ, T. MUELLER, H. WINKELMANN, AND W. R. FAHRNER. *Heterojunction emitter for crystalline silicon thin-film cells.* Proceedings of the 33rd IEEE Photovoltaic Specialists Conference (IEEE-PVSC-33), San Diego, USA (2008).

[4] W. R. FAHRNER, M. SCHERFF, S. SCHWERTHEIM, T. MUELLER, W. DUENGEN, B. STUETZEL, M. TROCHA, AND A. EBBERS. *Heterojunction solar cells made of CVS nanoparticles.* Proceedings of the 23rd European Photovoltaic Solar Energy Conference and Exhibition (EU-PVSEC-23), Valencia, Spain (2008).

[5] M. MUEHLBAUER, V. GAZUZ, R. AUER, T. MUELLER, AND W. R. FAHRNER. *Al/Si back contact with improved resistivity and contact resistance by an optimized RTP temperature-time profile.* Proceedings of the 33rd IEEE Photovoltaic Specialists Conference (IEEE-PVSC-33), San Diego, USA (2008).

[6] S. SCHWERTHEIM, M. LEINHOS, T. MUELLER, H. C. NEITZERT, AND W. R. FAHRNER. *PEDOT with carbon nanotubes as a replacement for the transparent conductive coating (ITO) of a heterojunction solar cell.* Proceedings of the 33rd IEEE Photovoltaic Specialists Conference (IEEE-PVSC-33), San Diego, USA (2008).

[7] S. SCHWERTHEIM, M. SCHERFF, T. MUELLER, W. R. FAHRNER, AND
    H. NEITZERT. *Lead-free electrical conductive adhesives for solar cell in-
    terconnectors.* Proceedings of the 33rd IEEE Photovoltaic Specialists Con-
    ference (IEEE-PVSC-33), San Diego, USA (2008).

[8] W. R. FAHRNER, M. SCHERFF, T. MUELLER, AND S. SCHWERTHEIM. *Evalua-
    tion of interface defects in a-Si:H/mc-Si heterojunction solar cells.* Techni-
    cal Digest of the International Photovoltaic Science and Engineering Con-
    ference (Asia-PVSEC-17), Fukuoka, Japan, pp. 4P–P2–23 (2007).

[9] W. R. FAHRNER, M. SCHERFF, T. MUELLER, AND S. SCHWERTHEIM. *Simu-
    lation of interface states in a-Si:H/mc-Si heterojunction solar cells.* NREL
    17th Workshop on Crystalline Silicon Solar Cells and Modules: Materials
    and Processes, Vail, Colorada USA (2007).

[10] W. DUENGEN, R. JOB, Y. MA, Y. HUANG, T. MUELLER, W. R. FAHRNER, L. O.
     KELLER, J. T. HORSTMANN, AND H. FIEDLER. *Thermal evolution of hydrogen
     related defects in hydrogen implanted Czochralski silicon investigated by
     Raman spectroscopy and atomic force microscopy.* Journal of Applied
     Physics, vol. 100, p. 034911 (2006).

[11] W. DUENGEN, R. JOB, T. MUELLER, Y. MA, W. R. FAHRNER, L. O. KELLER,
     J. T. HORSTMANN, AND H. FIEDLER. *Blistering of implanted crystalline
     silicon plasma hydrogenation by Raman spectroscopy.* Journal of Applied
     Physics, vol. 100, p. 124906 (2006).

[12] W. R. FAHRNER, T. MUELLER, M. SCHERFF, AND D. KNOENER. *Interface
     states of heterojunction solar cells.* Proceedings of the 4th IEEE World Con-
     ference on Photovoltaic Energy Conversion (WCPEC-4), Waikoloa, Hawaii,
     USA (2006).

[13] W. R. FAHRNER, T. MUELLER, M. SCHERFF, D. KNOENER, AND H. C.
     NEITZERT. *Defect states at the a-Si:H/c-Si interface of heterojunction solar
     cells.* Tech. rep., Jahrestagung PV-Uni-Netz, Hamburg, Germany (2006).

[14] H.-C. NEITZERT, M. FERRARA, T. MUELLER, M. SCHERFF, AND W. R.
     FAHRNER. *Bulk and interface degradation of amorphous silicon / crys-*

talline silicon heterojunction solar cells under proton irradiation. Proceedings of the 4th IEEE World Conference on Photovoltaic Energy Conversion (WCPEC-4), Waikoloa, Hawaii, USA (2006).

[15] M. SCHERFF, Y. MA, W. DUENGEN, T. MUELLER, AND W. R. FAHRNER. *Increased efficiencies in a-Si:H(n) / CZ-Si(p) heterojunction solar cells due to gradient doping by thermal donors.* Proceedings of the 21st European Photovoltaic Solar Energy Conference and Exhibition (EU-PVSEC-21), Dresden, Germany (2006).

[16] M. SCHERFF, S. SCHWERTHEIM, Y. MA, T. MUELLER, AND W. R. FAHRNER. *10x10 cm² HIT solar cells contacted with lead-free electrical conductive adhesives to solar cell interconnectors.* Proceedings of the 4th IEEE World Conference on Photovoltaic Energy Conversion (WCPEC-4), Waikoloa, Hawaii, USA (2006).

[17] W. R. FAHRNER, R. GOESSE, M. SCHERFF, T. MUELLER, M. FERRARA, AND H. C. NEITZERT. *Admittance measurements on a-Si/c-Si heterojunction solar cells.* Journal of the Electrochemical Society, vol. 152 (11), pp. 819–823 (2005).

[18] M. SCHERFF, H. WINDGASSEN, Y. MA, H. STIEBIG, T. MUELLER, AND W. R. FAHRNER. *Electrical properties of 10x10 cm² HIT solar cells with screen-printed metal grids manufactured on textured p-type mc-Si substrates.* Proceedings of the 15th International Photovoltaic Science and Engineering Conference (Asia-PVSEC-15), Shanghai (2005).

ACKNOWLEDGMENTS

I always read the acknowledgments page when I buy a new book. I like to know a little more about who helped shape a book and how it came to be, so I hope you'll take a second and read a little about the excellent team I got to work with to make the book that you're holding a reality.

I would like to express my respect and most sincere gratitude to my supervisor *Prof. Dr. Wolfgang R. Fahrner*. His guidance and encouragement, and advice through all this years is highly appreciated. He always has been extremely generous with his time and knowledge at all stages of my work and allowed me great freedom in research.

I have to acknowledge *Prof. Dr. Heinrich Christoph Neitzert* for willingly accepting to evaluate this thesis and be a member of the graduation committee.

I am greatly appreciative to all members of the group at LGBE for their long-lasting support. Especially I am obliged to *Boguslaw Wdowiak*, for helpful discussions and for maintenance of the PV labs at the LGBE, they are a tremendous place to work and study. I am deeply indebted to *Katrina Meusinger*, offering suggestions and encouragement, and providing me with millions of samples for a successfully completion of this work - if nothing else, then for bringing the most awesome refreshments. A smarty-pants badge of honor to *Stefan Schwertheim* - all of his continued support and urging is deeply appreciated. To those who joined the party late, but accomplished great help by taking numerous measurements: *Ruediger Klimkeit*, *Oliver Grewe*, and *Marcel Leinhos*. I am greatly appreciative for that. To *Barbara Kleine* who has made this experience all the more enjoyable. I owe a debt to those, who left the ship early, but supported me in my research work and helped to get off on the right foot: *Maximilian Scherff, Yue Ma, Yuelong*

Huang, Reinhart Job, and *Wolfgang Duengen* – I want to thank all of them for theirs help, support, interest and valuable hints.

A special thanks goes to my parents, *Dieter and Marianne Mueller,* who have demonstrated their own loving patience with their still-growing son through the years and who enabled me in my early years to scout the world and to explore my interests with remarkable patience.

Last but foremost I want to thank my passionate early reader: my wife, *Natalie Mueller,* Dr. rer. nat., an accomplished physicist who has shown so much loving patience with her husband's interests. I am very proud of her own abilities and have come to appreciate deeply how much human values are important. This thesis may have not been possible for me without taking a leaf out of her book. If I'd wrote down everything I ever wanted, I would not have believed I could meet someone better!

Thomas Mueller
April 2009

CURRICULUM VITAE

Personal Details

Date / Place of birth	June 8th, 1978 / Hamm, Germany
Family status	married
Citizenship	german

Education

University of Hagen, Department of Mathematics and Information Technology, Electronic Devices (LGBE), Hagen, Germany

Postgraduate Research Assistant   2004 – 2009
  Includes current Ph.D. research, Ph.D. level coursework and
  research/consulting projects.

Center for Advanced Microstructures & Devices (CAMD), Louisiana State University, Baton Rouge, Louisiana, USA

Research Assistant   2003

University of Dortmund, Department of Electrical Engineering, Dortmund, Germany

Graduate Student   1998 – 2004
  Dipl.-Ing. (M.Sc.), Electrical Engineering, Institute Microstructure
  Technology (MST), 2004.

Galilei-Gymnasium, Hamm, Germany

Secondary School   1984 – 1997
  Graduated with Abitur (German University Entrance Qualification), 1997.